Airplane Maintenance and Repair

Other McGraw-Hill Aviation Titles

Aircraft Systems by *David A. Lombardo*

Advanced Aircraft Systems by *David A. Lombardo*

Standard Aircraft Handbook, Fifth Edition
by *Larry Reithmaier*

How to Make Your Airplane Last Forever
by *Mary Woodhouse and Scott Gifford*

Airplane Ownership by *Ronald J. Wanttaja*

The Right Seat Handbook: A White-Knuckle Flier's
Guide to Light Planes by *Charles F. Spence*

Just Plane Smart: Activities for Kids
in the Sky by *Ed Sobey*

Airplane Maintenance and Repair

A Manual for Owners, Builders, Technicians, and Pilots

Douglas S. Carmody

McGraw-Hill

New York San Francisco Washington, D.C. Auckland Bogotá
Caracas Lisbon London Madrid Mexico City Milan
Montreal New Delhi San Juan Singapore
Sydney Tokyo Toronto

Library of Congress Cataloging-in-Publication Data

Carmody, Douglas S.
Airplane maintenance and repair : a manual for owners, builders, technicians, and pilots / Douglas S. Carmody.
p. cm.
Includes index.
ISBN 0-07-011937-6 (pbk.)
1. Airplanes—Maintenance and repair—Handbooks, manuals, etc.
I. Title.
TL671.9.C33 1997 97-28746
629.134'6—dc21 CIP

McGraw-Hill

A Division of The **McGraw-Hill** *Companies*

1 2 3 4 5 6 7 8 9 0 FGR/FGR 9 0 2 1 0 9 8 7

ISBN 0-07-011937-6

The sponsoring editor for this book was Shelley Ingram Carr, the editing supervisor was Ruth W. Mannino, and the production supervisor was Sherri Souffrance. It was set in Garamond as a GEN1-5×8 by McGraw-Hill's Hightstown, N.J. desktop publishing department.

Printed and bound by Quebecor/Fairfield.

For Bonny, Caroline, and Mary Catherine

Contents

A mechanic inspects the brakes of a DC-10 at Honolulu International Airport.

Preface

One can argue that maintaining an aircraft is more expensive than flying it. It is true that good maintenance isn't cheap, but it can be affordable, especially if the owner/pilot is willing to become an informed consumer. This book will help you be an informed consumer and allow you to make maintenance decisions based on knowledge rather than solely on the aviation maintenance technician's diagnosis. This book will help you work in concert with your mechanic to produce less downtime and less expensive maintenance bills.

For the mechanically inclined, step-by-step guidelines indicate how to perform preventive maintenance. Before attempting to use the information presented in this book, assess your own level of mechanical ability. Even if you're all thumbs around tools, the information compiled in this book can help save you thousands of dollars in troubleshooting and repair work.

Remember, knowledge is power! During training at the airline I fly for, one of the most informative classes was conducted by a line mechanic. He came to the class and explained how an informed pilot could save the company literally hundreds of worker hours and thousands of dollars by being as precise as possible when writing up a maintenance problem. The more you know about your aircraft's systems, the better you'll be at identifying a small problem before it becomes a big one. As pilots, we often are not as well versed as mechanics are on the nuts and bolts operation of our aircraft, and this book is designed to inform you about the operating systems of your aircraft and about inventory maintenance. Listed at the back of each chapter is a guideline for preventive maintenance—exactly who can perform it, the standards that must be met while doing the maintenance, and who can return the aircraft to service.

Let's start by determining just what the Federal Aviation Administration (FAA) considers preventive maintenance. Preventive maintenance consists of those tasks specifically listed by Federal Aviation Regulation (FAR) 43, Appendix A, paragraph (c). (Those items appear in Chap. 8 of this book.) If the task isn't listed there, then it isn't considered preventive maintenance, and a licensed mechanic has to do the work. If your aircraft is operated under FAR 121, 135, or 127, all maintenance work must be accomplished by a certified mechanic regardless of whether you own the aircraft. The regulations go on to say that the holder of a pilot certificate issued under FAR 61 is authorized to perform any of those items listed as preventive maintenance. The certificate holder must have a private pilot license or higher, however. Keep in mind that different aircraft call for different procedures. Just because a certain operation is listed as preventive maintenance by FAR 43, don't assume that it is a simple operation to perform on your aircraft. Because of difference in aircraft, a function may be preventive maintenance on one aircraft and not on another. Owners and pilots must use their own good judgment in determining if a specific function can actually be classified as preventive maintenance.

The most important step in doing your own maintenance is determining if you have the knowledge and ability to complete the task safely. Take an honest look at your capabilities. Some people are just not mechanically inclined. In addition, some people don't have the correct tools to try to tackle all the items listed in FAR 43: Not only must you have the ability to accomplish a task, but by regulation you must have the correct tools and equipment. FAR 43.13 requires the use of "tools, equipment, and test apparatus necessary to assure completion of the work in accordance with accepted industry practices." Normally these are listed as part of any FAA-approved manufacturer's maintenance manual. If you don't have the maintenance manuals to your aircraft, try to get a set. The few dollars you spend on them will be saved hundreds of times over. Develop a good working relationship with your mechanic—be up front and honest about performing your own preventive maintenance. There will be times when you need to ask a licensed mechanic his or her opinion or advice regarding airworthiness issues. Don't expect the mechanic to work for free or to drop everything to go take a look at something on your airplane. Most mechanics will go out of their way to be helpful, but I still wouldn't ask to borrow too many tools!

Probably the quickest way to get an FAA inspector interested in your aircraft is to have evidence of obvious maintenance with no logbook entries. One pilot had three new tires on his Tri-Pacer and got extremely red-faced as he tried to explain to the FAA during a ramp check why the last logbook entry showing tire changes was over 5 years old! Remember, the job isn't finished until the paperwork is complete. Working on your own aircraft can add immeasurably to the joy and pleasure of owning your own aircraft. Like most worthwhile endeavors, if it's worth doing, it's worth doing right.

Acknowledgments

This book would not have been possible without the help and encouragement of Shirley Kümpf—thanks for the typing and retyping. Thanks also to my brother Raymond, a far better mechanic than I, for his expert advice and support.

The book is a compilation of information from many sources, including the FAA, industry, aircraft manufacturers, and line mechanics. For clarity, some FAA information has been included in its entirety.

Airplane Maintenance and Repair

1

Reciprocating Engine Operation and Repair

The only thing some pilots want to know about their airplane's engine is if it will start when they turn the key. Others want to know about every bore, every stroke, and every compression ratio. The vast majority of pilots are somewhere in between. They have picked up pearls of mechanical wisdom from flight instructors and other pilots. Some of the information is good, some of it is bad, but most pilot's knowledge of mechanical systems is incomplete, and ignorance about aircraft engines can hurt you, both physically as well as financially.

To understand today's modern reciprocating airplane engine, a little history lesson is in order. Charles Taylor built the first successful airplane engine in 1903. That engine flew on the Wright Flyer in December of that year. As airplane engines go, it wasn't much. It produced just 16 hp and weighed a whopping 180 lb. Like a red cape waved in front of a bull, that statistic has been challenging airplane engine manufacturers ever since. Weight per horsepower is their holy grail. Their goal is getting the most horsepower out of the lightest engine possible. To compare how far they've come, the Continental 0-200 also weighs 180 lb but delivers a 100 hp, giving it a weight-to-power ratio of 1.7 to 1. (See Fig. 1-1.)

Manufacturers have coined a term used to measure the durability and reliability of their engines: *TBO* is *t*ime *b*etween *o*verhauls. Unless the aircraft is operated commercially (i.e., under FAR 135) it is just a factory recommended time frame. There is no legal obligation to overhaul the engine once that time has accumulated. However, to put TBO in some perspective, an average overhaul period of 2000 hours would equate to approximately 250,000 miles on the family car. That's something to think about when considering buying a high-time-used airplane or loading the family up for

1-1 *Continental 0-200 engine. (Courtesy of Michael J. Kroes and Thomas A. Wild, Aircraft Powerplants, McGraw-Hill, 1995)*

a night cross-country with an engine over its TBO. The biggest TBO killers are neglect and infrequent use. Letting an engine sit for a long period is actually harder on it than long flights at cruise power. Engines that sit idle for long periods accumulate contaminants that form inside the engine. To dissipate these harmful contaminants, the engine should be run at least weekly and allowed to reach its operating temperatures throughout. Flying the aircraft is just about the only way to get the temperatures up sufficiently to do any good. Sitting on the ramp and letting it idle rarely will get the internal temperatures high enough.

Engine Construction and Operation

Current engine manufacturers use letters to indicate the type and characteristics of the engine followed by numbers to indicate displacement. The following nomenclature will help identify any engine:

- L Leftward rotation used on counter-rotating propellers
- G Geared—the propeller is turned through a reduction gear box
- T Turbocharged
- S Supercharged—the engine is equipped with an engine on turbine driven supercharger
- V Vertical installation (for helicopters)
- H Horizontal installation (for helicopters)
- O Opposed cylinders
- I Fuel injection system
- A Aerobatic—the fuel and oil systems are designed for inverted flight

As an example, a *Lycoming 0-320-E2D* engine indicates an opposed 320 in^3 displacement *E* type crankcase; *2* indicates the type of nose section used, and the *D* refers to the type of accessory drive utilized.

Crankcase

The crankcase is the foundation of any engine (see Fig. 1-2), housing the various parts that, working together, make up an engine. Most aircraft engine crankcases are manufactured out of an aluminum alloy. This provides the crankcase with strength while maintaining light weight. Unfortunately, cast aluminum is more subject to cracking than cast iron, but the weight of a cast iron crankcase would be prohibitive in aviation applications. Using a sandcast or permanent molding cast, the crankcase is manufactured by the engine builders or an outside vendor to specific standards set by the engine's manufacturer. The crankcase then goes through a final machining process, followed by the installation of numerous studs along the edges of the crankcase. These studs are used as

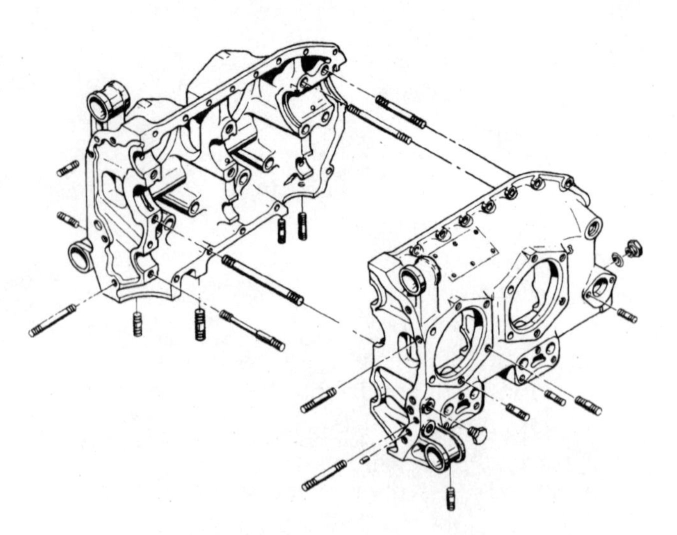

1-2 *Crankcase for a four-cylinder opposed engine. (Courtesy of Michael J. Kroes and Thomas A. Wild, Aircraft Powerplants, McGraw-Hill, 1995)*

pins to hold the two halves of the engine together. Mounting pads are machined into the crankcase to hold the cylinders, oil sump, and other accessories. Galleries and oil passages are drilled into the crankcase to allow the oil to circulate freely and to lubricate the crankshaft bearings, the camshaft bearing, and the myriad of other moving parts. The crankcase is also vented to the outside air through a breather tube. Finally the main bearing bores are machined into the case. Once assembled, the crankcase is held together by a series of studs and sealants, normally without the benefit of a gasket.

Crankshaft

The back-and-forth motion of the pistons is converted to a rotary motion by the crankshaft. Unlike automobile crankshafts, which are made of cast iron, an airplane's crankshaft is a forged steel alloy. Usually the crankshaft is made of chromium-nickel-molybdenum steel, an extremely hard and durable metal. Think of the crankshaft as the backbone of an engine. It's strong, long, and, believe it or not, flexible. Of course, the amount of give or flex in a crankshaft depends on how long it is. A large six-cylinder engine such as a Continental IO-520 has a long crankshaft and has to use counterweights to dampen vibrations. Shorter crankshafts are stiffer and may not need any counterweights. Of course, long or

1-3 *Engine crankshaft. (Courtesy of Michael J. Kroes and Thomas A. Wild, Aircraft Powerplants, McGraw-Hill, 1995)*

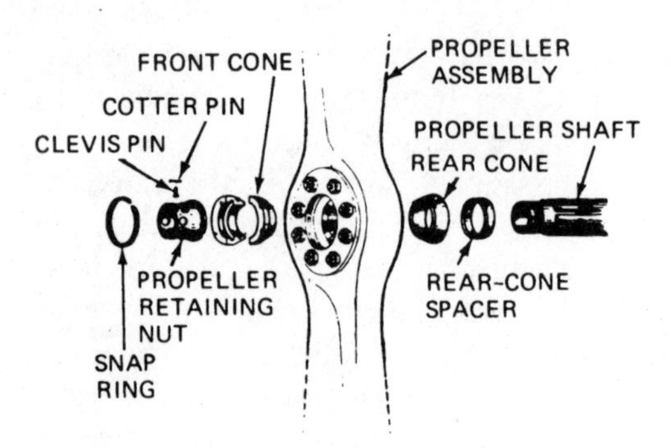

1-4 *A flange-type shaft used to attach the propeller. (Courtesy of Michael J. Kroes and Thomas A. Wild, Aircraft Powerplants, McGraw-Hill, 1995)*

short, the crankshaft has to fit inside the crankcase. Once inside the crankcase, the crankshaft is supported by the main bearing journal. (See Fig. 1-3.) This journal holds the main bearing which serves as the center of rotation of the crankshaft. Every airplane engine has at least two main bearing journals. Some have more, but none have less. On direct drive engines the propeller shaft is also an integral part of the crankshaft, and this shaft protrudes from the front of the crankcase. Fixed to the propeller shaft is a flange type shaft used to attach the propeller. Normally, six high-strength bolts are used to secure the propeller to the flange. (See Fig. 1-4.) While flying, air loads are transmitted through the propeller shaft right into the engine. To help dampen these loads, a main bearing and a thrust bearing are incorporated in the nose section of the engine. On some engines the main bearing is flanged, to allow it to serve as both a main bearing and a thrust bearing.

Connecting Rod

The connecting rod is the mechanism that transfers the back-and-forth action of the piston to the rotary

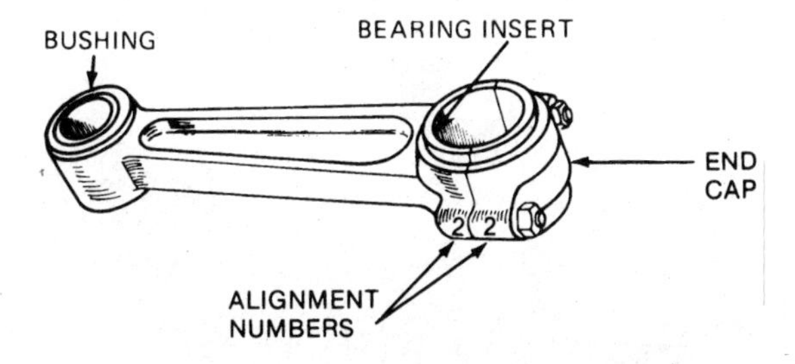

1-5 *A connecting rod. (Courtesy of Michael J. Kroes and Thomas A. Wild, Aircraft Powerplants, McGraw-Hill, 1995)*

action of the crankshaft. (See Fig. 1-5.) This, in turn, is what drives the propeller. The part of the connecting rod that actually attaches to the crankshaft is called the *crankpin end,* and the part of the connecting rod which attaches to the piston is called the *piston pin end.* During any inspection, repair, or overhaul, it is important to replace the connecting rod in the same cylinder and in the same position prior to its removal. This maintains a proper fit and balance.

The Piston

A piston is like a plunger. (See Fig. 1-6.) It is moved up and down in the cylinder by the expansion of the burning fuel air mixture. This movement is transmitted through the connecting rods to the crankshaft, and ultimately to the propeller. This is where the term *four-stroke engine,* originated. Here's what happens:

1. As the piston moves down toward the crankshaft, it draws in the fuel air mixture (intake).
2. As the piston moves up toward the top of the cylinder, it compresses the fuel charge (compression).
3. Ignition occurs, and the rapid burning of the fuel air mixture drives the piston back toward the crankshaft (power).

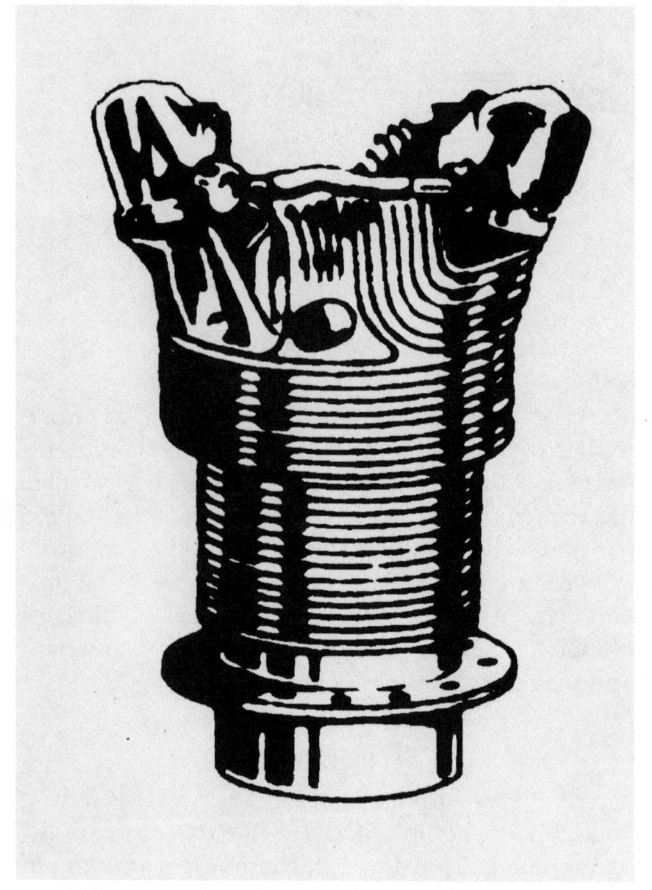

1-6 *A piston.*

4. The piston makes the fourth stroke as it moves back up to the top of the cylinder forcing the exhaust out of the combustion chamber (exhaust).

Of course, the pistons have to be extremely durable to handle all that stress and heat. A piston is usually made of aluminum alloys to take advantage of their light weight and excellent heat conductivity. The top of the piston is called the *head,* and the sides are

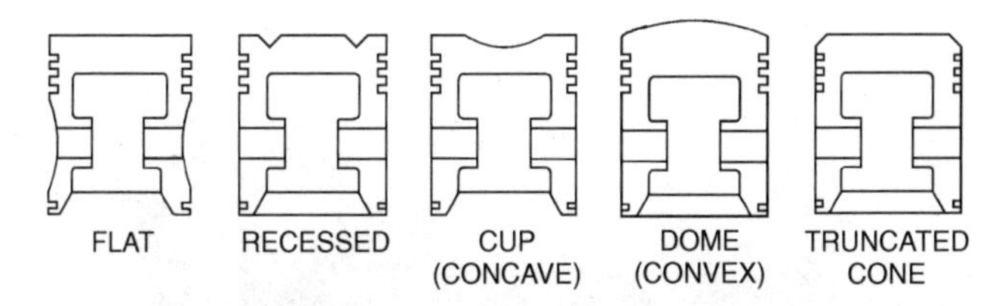

1-7 *Types of piston heads. (Courtesy of Michael J. Kroes and Thomas A. Wild, Aircraft Powerplants, McGraw-Hill, 1995)*

called the *skirts*. Underneath the piston are ribs which expand the surface area. This expanded area provides a greater surface for the lubricating oil to disperse the enormous heat generated at the piston head. Pistons are classified according to the shape of the piston head. There are flat head pistons (the most common) as well as concave, recessed, domed, or convex and truncated cone. (See Fig. 1-7.) Engines that are basically the same can have varied horsepower ratings merely by changing the type of piston.

Cylinders

Inside the cylinder, the burning of the fuel is converted into mechanical energy. (See Fig. 1-8.) In addition to converting the fuel air mixture into power, the cylinders are also designed to dissipate heat. A modern cylinder has eight parts. Starting at the bottom of the cylinder is an area called the *skirt*. This part of the cylinder extends beyond the flange and actually into the engine crankcase. This design utilizes shorter connecting rods and allows the manufacture to build a more compact engine. Connected to the skirt is the flange. This predrilled attachment plate is bolted directly to the crankcase. The heart of the cylinder is the barrel. The cylinder barrel is where the piston moves up and down or *reciprocates*. (Hence the term *reciprocating engine*.) The barrel itself must be strong and lightweight, and the inside of the cylinder

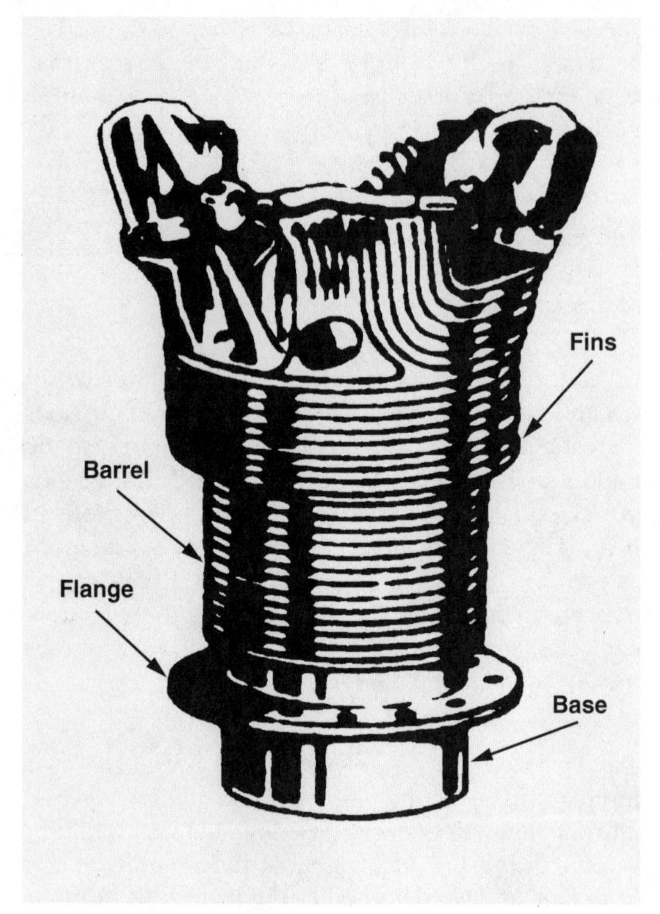

1-8 *Cylinder assembly.*

can be neither too smooth nor too rough. A too-rough surface will damage the piston rings and the cylinder walls. A too-smooth surface will not hold enough oil against the walls during the engine's break-in period. Think first about how often that piston moves up and down that barrel during the course of a single flight and then consider how often the piston reciprocates during the lifetime of the engine. That constant wear has to be controlled, and the most common way to harden the cylinder walls is through nitriding or chromium plating. Nitriding is a process, not a coating. The cylinder is baked in an atmosphere of ammonia gas; this gas breaks down into components of nitrogen and hydrogen, and the nitrogen bonds with the steel in the cylinder, creating a hard, wear-resistant surface. Unfortunately, nitrided cylinders tend to rust easily if they are not kept lightly coated with oil. When storing an aircraft for a few months (like over the winter), it's an excellent idea to use a preservative oil. A nitrided cylinder can be identified by the blue paint around the base of the cylinders. (See Fig. 1-8.)

Chromium plating is usually done during an engine overhaul. When a cylinder is "chromed," it undergoes a process where a coating of chromium is adhered electrolytically to the cylinder walls. The chromium process has many advantages over nitrided or steel cylinders. Chromium cylinders are much less prone to rusting, because oil sticks to the chromium better than it does to steel, and has a very low coefficient of friction and excellent wear properties. Chromed cylinders are easily identified by an orange band painted around the cylinders' base. (See Fig. 1-8.)

Combustion takes place at the cylinder head, which contains the intake and exhaust valves.Located on top of the cylinder head is the rocker bosses. These are shaft guides for the rocker arms. The spark plug holes are located at strategically placed positions on the cylinder head to ensure an optimum burning pattern. Some of these holes have reinforced steel inserts called *HeliCoils*. These allow for fast and easy replacement of the threads if stripped or otherwise damaged during spark plug changes. The cylinders may also be choke-bored. This is a slight taper built into the cylinder itself. The bore at the cylinder head is slightly smaller than the bore at the cylinder's skirt. This allows for expansion once the engine reaches operating temperatures. Once warmed up, the cylinder will expand to the point that the bore is nearly consistent throughout the cylinder. Although a cylinder would seem to be one piece of steel, this is not usually the case. During manufacturing most cylinder heads are first heated to a very high temperature, then they are screwed onto the barrel of the cylinder. During the manufacturing process, the barrel of the cylinder has been chilled to shrink it. After assembly, the two pieces return to a normal temperature and size, forming a gas tight seal.

Cooling fins allow for the cylinders to dissipate enormous amounts of heat built up while operating. (See Fig. 1-9.) These fins are cast or machined on to the outside of the cylinder barrel and head. They are set in a pattern designed to make the most efficient use of the ambient air for cooling purposes. Note that the intake valve has no cooling fins. This is because the incoming fuel air mixture acts as a coolant and eliminates the need for any cooling fins. It also makes it very easy to identify the intake or exhaust side of a cylinder.

Spark Plugs

A spark plug provides the spark at the appropriate time, causing the fuel air mixture in the cylinder to ignite. A spark plug needs to resist fouling from the combustion by-products, and the electrode itself has to be able to withstand the repeated arcing of the spark without undue wear. Although a spark plug seems a simple device, it's actually a very complex component. (See Fig. 1-10.)

Size

Aircraft spark plugs are available in two sizes of threads, 14 mm or 18 mm. The most commonly used spark plug is the 18 mm variety. The thread length of a spark plug is called its *reach*. A plug with a reach of ½ in would be classified as a short reach plug, one

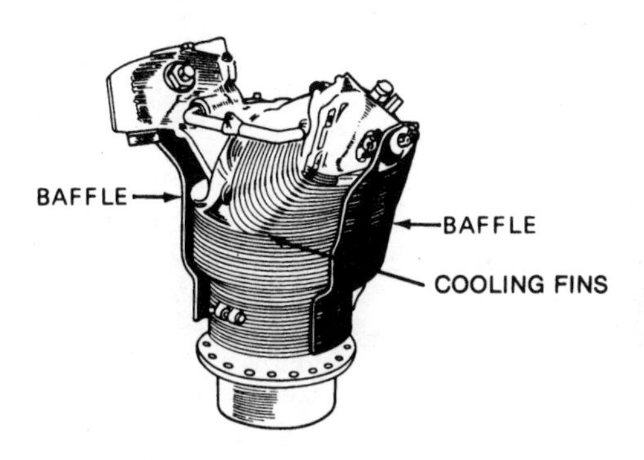

1-9 *Cylinder with cooling baffles. (Courtesy of Michael J. Kroes and Thomas A. Wild, Aircraft Powerplants, McGraw-Hill, 1995)*

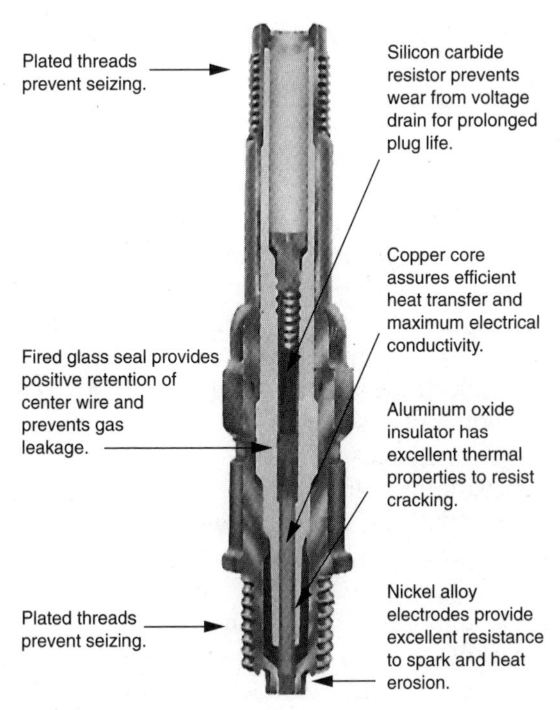

1-10 *Shielded spark plug. (Courtesy of Cooper Industries, Inc.)*

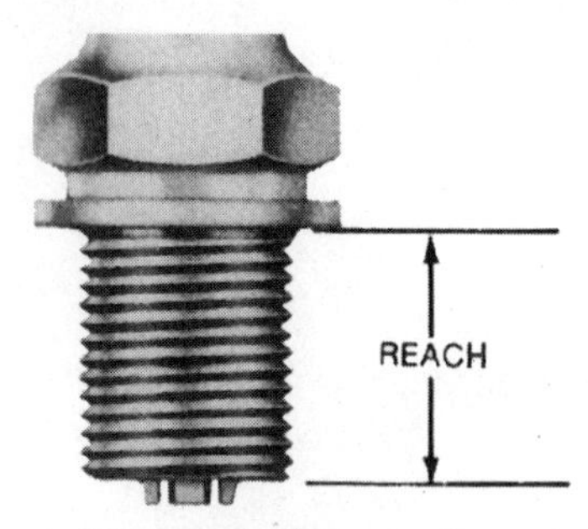

1-11 *Spark plug reach. (Courtesy of Michael J. Kroes and Thomas A. Wild, Aircraft Powerplants, McGraw-Hill, 1995)*

of $^{3}/_{16}$ in would be considered a long-reach plug. (See Fig. 1-11.)

Electrodes

The firing end of a spark plug has three prongs. The center prong is an electrode. It's from here that the spark will arc across to another prong. This arc creates the heat necessary for igniting the fuel air mixture. There are two types of electrodes, massive and fine wire. (See Fig. 1-12.) The massive electrode is a copper material encased in a nickel sheath. This design allows for a maximum transfer of heat between the electrode and a prong and resists the buildup of hot spots. A fine wire electrode, as the name implies, consists of two very fine wires made of iridium or platinum. Here, the electrode can arc across the gap utilizing a much lower voltage. The *heat range* of the spark plug refers to its ability to dissipate heat away from the tip of the plug and into the shell of the spark plug. A hot plug has a long way for the heat to travel from its tip, and these are used in relatively cool combustion chambers like those found in low-combustion engines. A cold plug, of course, would be used in the opposite type of engine, a high

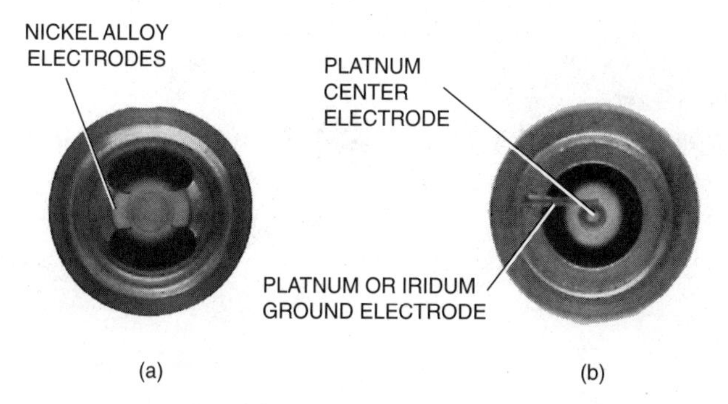

1-12 *Example of (a) massive and (b) fine wire spark plugs. (Courtesy of Cooper Industries, Inc.)*

compression, hot burning engine. Care must be taken when choosing a spark plug, to ensure that it burns hot enough to reduce fouling but doesn't burn so hot that it leads to preignition. As always, be sure to follow the manufacturers' recommendations.

Inspection

A good mechanic can tell a lot about the condition of an engine by "reading" the spark plugs. Spark plug wear reveals a tremendous amount of information about what is occurring to the engine internally. When the spark plugs are removed from the engine, always put them in a rack that will easily identify what cylinder they belong in. It's very frustrating finding a serious problem on a spark plug and not remembering which cylinder is the culprit. If the plugs have been removed and they show a dull brown deposit on the insulator, then the cylinders have been operating properly. You can clean them, regap them, and replace them till the next inspection.

Damaged or Worn Plugs

To determine if the spark plugs are worn out, use this simple guide. On fine wire plugs, you can continue to

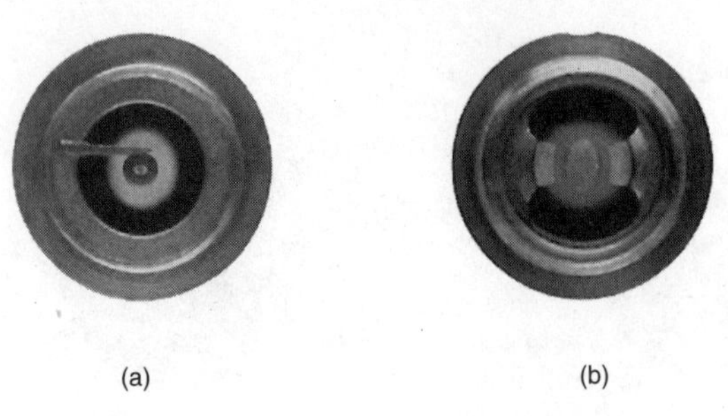

1-13 *Examples of acceptable wear: (a) fine wire; (b) two-prong ground electrodes. (Courtesy of Cooper Industries, Inc.)*

use the plugs until half of the original size of the electrode or ground prongs have been worn away. (See Fig. 1-13.) Massive electrode plugs can be regapped and reused until more than half of the nickel sheath has worn away. Normally the sheath is 0.003 mm thick and can be reused until the plug gets down to 0.015 mm. If the plugs are borderline, discard them and replace them with new ones. If, upon inspection of the plugs, severe erosion is found, be sure to investigate prior to flying the aircraft again. The most likely culprit is an induction system leak, and the piston head and cylinder head should be checked for damage. This can be accomplished by having the engine borescoped.

Lead Deposits and Fouling

Deposits of lead and plug fouling probably have occurred to everyone flying with avgas that incorporates tetraethyl lead as an antidetonation agent. Too much fuel or a spark plug running too cold can lead to a buildup of lead deposits in the cavity between the conductor and the shell. Lead deposits are easy to identify; they look like little beads of molten metal clinging to the inside of the spark plug. Fine wire plugs with platinum electrodes may be unsalvageable if the lead buildup is too great.

The following steps should be followed to clean used spark plugs: After removing the spark plugs from the engine, clean and degrease them with Varsol or some other cleaning solvent. Exercise care if cleaning plugs with unleaded gasoline to prevent a fire. If you use a solvent, be careful not to get it inside the terminal end of the plug. Next, wire-brush the threads to remove any built-up dirt. Many pilots have access through a friendly mechanic or shop to an abrasive blast cleaning machine. Using glass beads, these machines use air pressure to blast the electrodes and spark plug cavity to clean out any buildups. During a normal 100 hours of operation, the average spark plugs are going to fire approximately 8 million times. That in itself can wear away a significant portion of the electrodes. An improper or too-long blast from an abrasive cleaner can easily wear away two to three times that much electrode. Make sure to follow the blaster manufacturer's directions. After cleaning the firing end of the plug, clean the terminal end. It should be cleaned with a light abrasive cleanser and a swab. After cleaning, flush the terminal end with alcohol or acetone to remove all traces of the cleanser. Don't flush the plug with leaded gasoline, which is a great conductor and could cause the spark plug to misfire. When gapping the plugs prior to replacing them, "eyeballing" is not a recommended procedure. (See Fig. 1-14.) Using anything other than the spark plug manufacturer's procedures will result in shorter spark plug life. When gapping a massive electrode plug, never try to adjust the electrode. This will damage the ceramic surrounding the electrode and may destroy the spark plug. When adjusting the ground electrodes, always use the proper tools. Make sure that the ground electrodes are parallel to the center electrode, otherwise the edge of the center electrode

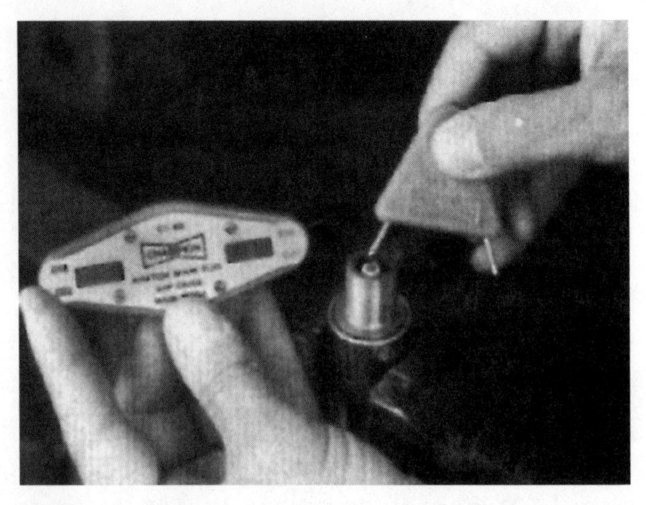

1-14 *Setting the gap on a fine wire spark plug. (Courtesy of Cooper Industries, Inc.)*

will wear away and widen the gap resulting in an incorrect setting. Always use a wire type gap gauge when checking or adjusting the gap on the spark plugs. Fine wire plugs are easier to adjust, but use caution because the wires are brittle and can snap off. When replacing the plugs, rotate them to the next cylinder in the firing order and change the bottom plugs to the top and the top to the bottom. Make sure that a new gasket is used when reinstalling the plugs. (See preventive maintenance procedures at the end of this chapter.) Put the gasket on the plug, then slip the plug into the hole. Thread it hand tight, being careful not to cross-thread it. Tighten the plug as far as possible by hand, utilizing a spark plug socket and torque wrench to finish the job. Tighten the spark plug to the engine manufacturer's recommended torque values. This is very important to get a good seal. Too much torque can damage the cylinder or stretch the spark plug to the point that it's damaged. Too little torque and hot gases can blow by the threads.

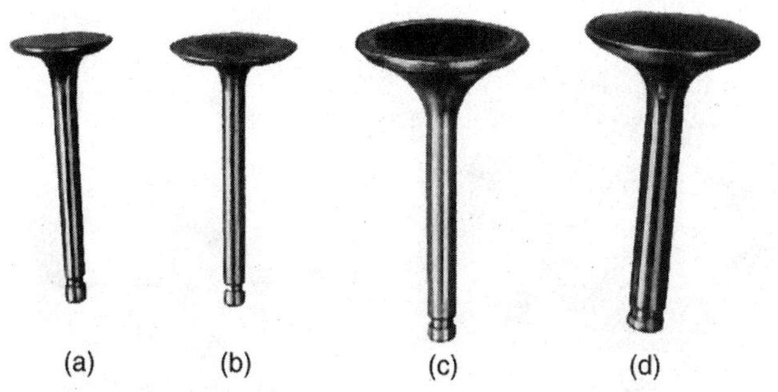

1-15 *Example of valve types: (a) flathead; (b) semitulip; (c) tulip; (d) mushroom. (Courtesy of Michael J. Kroes and Thomas A. Wild, Aircraft Powerplants, McGraw-Hill, 1995)*

Valves

A valve controls the flow of a liquid or a gas. (See Fig. 1-15.) In an aircraft engine, the valve opens and closes in two areas, called *ports*. One is the intake port, the other is the exhaust port. Each cylinder has one of each valve, for a total of two valves per cylinder. The intake valve lets in the fuel and air mixture necessary for combustion, and the exhaust valve allows the burned gases to be discharged.

Exhaust Valves

The exhaust valves are subject to high temperatures and have been designed to dissipate that heat as rapidly as possible. (See Fig. 1-16.) Some designs incorporate a hollow valve stem that is partially filled with sodium. Because sodium melts at approximately 200°F, the fluid flows back and forth in the valve stem, absorbing some of the heat and transferring it directly to the cylinder head. The cylinder head dissipates the heat through the cooling fins located on the outside of the cylinder. Of course, not all valves are sodium filled. In fact, Continental engines use a solid valve. Although its heat dissipation properties are not as

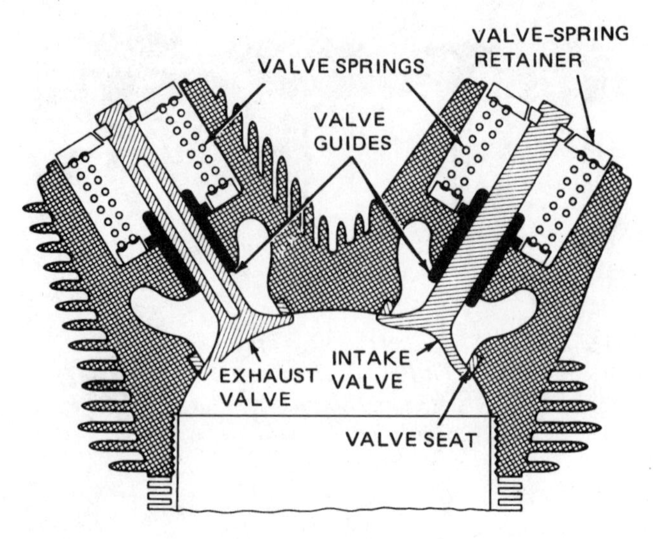

1-16 *Example of valve placement within a cylinder. (Courtesy of Michael J. Kroes and Thomas A. Wild, Aircraft Powerplants, McGraw-Hill, 1995)*

good as the sodium-filled valves, it is quite a bit less expensive.

Valve Guides

As their name implies, valve guides are used to support and position the stem of the valves. In order to obtain a tight fit, the cylinder head is heated during manufacturing to expand the holes where the valve guides are going to be placed. The guides are then pressed into position, and as the cylinder head cools down, the guide is held firmly in place.

Valve Seats

The materials used to make modern aircraft engine cylinders are not hard enough to withstand the constant hammering that occurs as the valves open and close. To remedy this, a valve seat is installed in the cylinder head. This seat is usually made of a forged aluminum

1-17 *Camshaft for a six-cylinder opposed engine. (Courtesy of Michael J. Kroes and Thomas A. Wild, Aircraft Powerplants, McGraw-Hill, 1995)*

bronze for the intake valve and a harder chromium-molybdenum steel for the exhaust valve seat.

Cam Shaft

Actually, the cam shaft should be called a *valve operating device.* (See Fig. 1-17.) A camshaft is just a metal rod with a number of strategically positioned lobes on it. These lobes operate all the exhaust and intake valves in the engine. As the crankshaft turns, so does the camshaft, only at a much slower rate, about one-half the speed of the crankshaft. This is appropriate, since the valves open and close only once each cycle, whereas the crankshaft makes two revolutions per cycle. As the camshaft turns, the lobe encounters a hydraulic lifter. This valve lifter engages a pushrod which is connected to a rocker arm. As the name implies, this rocker arm pushes down on the tip of the valve stem and overcomes the valve spring tension. As this occurs, the valve opens. Meanwhile, down on the camshaft, the lobe has rotated off the hydraulic lifter, allowing the valve spring tension to snap the valve shut.

Accessory Section

Usually located on the back of the engine, the accessory section provides a mounting area for engine driven accessories—things like magnetos, fuel pumps, or starters. Of course, the engine layout and manufacture determine the size and shape of the accessory section and also what items are located on it. (See Fig. 1-18.)

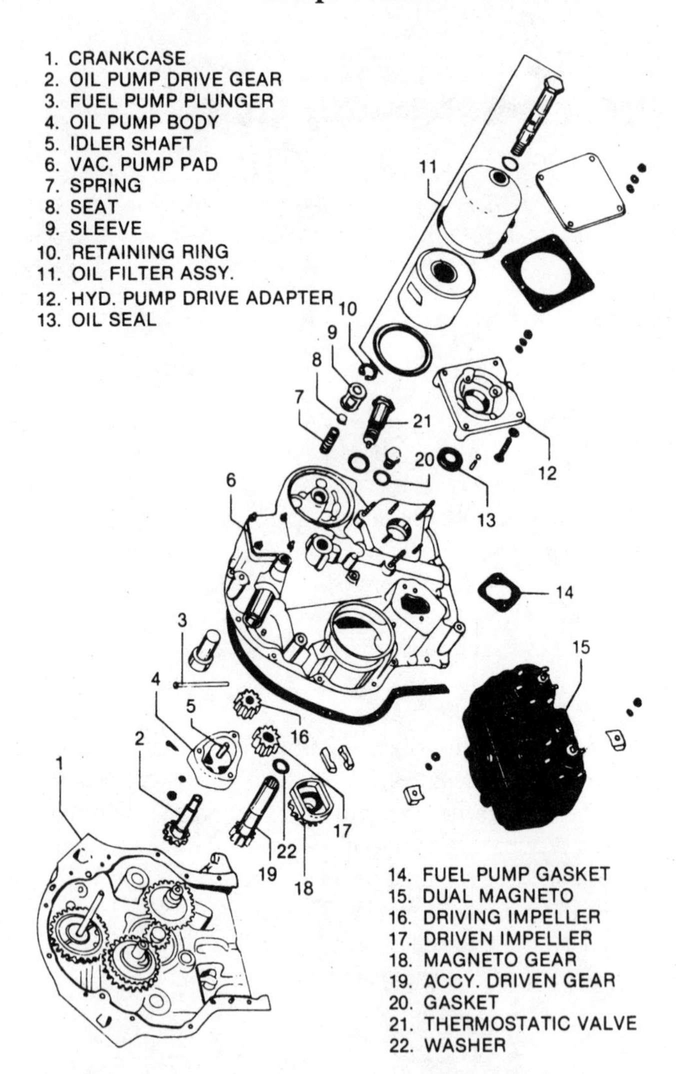

1-18 *Accessory case for a six-cylinder engine. (Property of Textron Lycoming, 652 Oliver St., Williamsburg, Part 17708. This is only an illustration and should not be used for actual maintenance).*

The Four-Stroke Engine

To fully comprehend what is occurring inside the engine, it's important to understand a few terms:

- *Cycle.* Like the four seasons, the engine goes through a cycle. A cycle is a complete series of events returning to the original position upon completion. It then starts all over again. Most piston aircraft engines operate on a four-stroke, five-event cycle developed by August Otto and consequently known as the Otto cycle. (See Fig. 1-19.)
- *Internal combustion engine.* Certain events occur inside the engine at certain times. Combustion occurs inside the engine rather than externally.
- *Stroke.* This is the distance the piston travels. For each turn of the crankshaft, the piston will travel two strokes: one up, one down. The limits of these strokes are *top dead center.*
- *Dead center.* The maximum for upward movement and *bottom dead center* for the maximum distance the piston will travel on the downward stroke.
- *Compression ratio.* This is the ratio of the volume of space in the cylinder when the piston is at bottom dead center to the volume of space in the cylinder when the piston is at top dead center. (See Fig. 1-20.) As an example, assume the volume of space in a cylinder is 140 in^3 when the piston is at the bottom of its stroke. The volume remaining in the cylinder is 20 in^3 at the top of the piston's stroke. So the ratio is 140/20. Now divide 140 by 20 to get a 7:1 compression ratio.

Remember earlier, we reviewed the four strokes the engine goes through while operating: (1) intake,

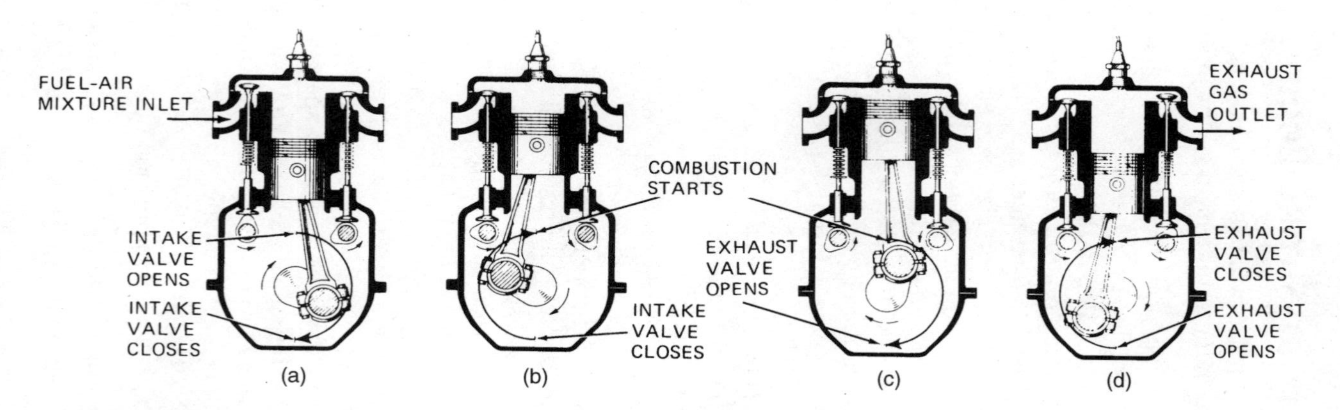

1-19 *The Otto cycle: (a) intake stroke; (b) compression stroke; (c) power stroke; (d) exhaust stroke. (Courtesy of Michael J. Kroes and Thomas A. Wild, Aircraft Powerplants, McGraw-Hill, 1995)*

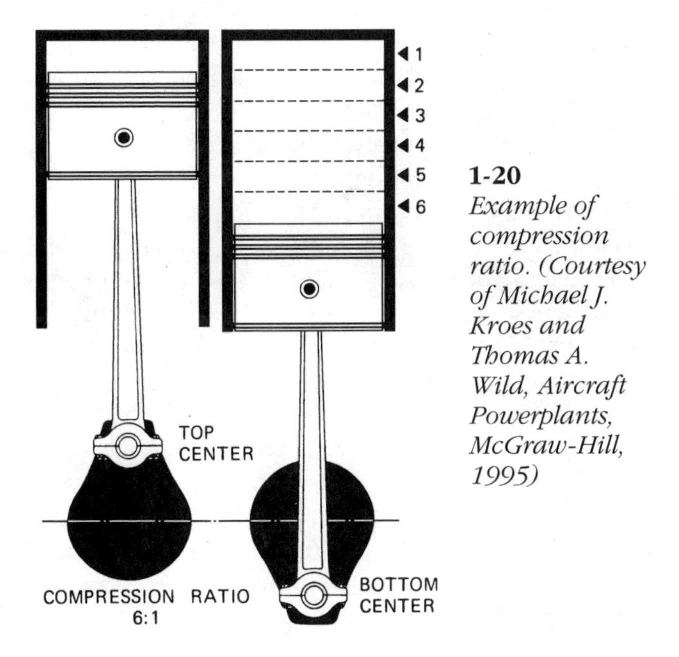

1-20
Example of compression ratio. (Courtesy of Michael J. Kroes and Thomas A. Wild, Aircraft Powerplants, McGraw-Hill, 1995)

(2) compression, (3) power, and (4) exhaust. By delving a bit deeper, it's possible to see what really happens inside a typical reciprocating engine: Starting with the *intake* stroke, the piston will be at top dead center. The intake valve will be open, and the exhaust valve will be shut. As the piston moves down the cylinder, a vacuum or negative pressure, is created in the cylinder. This vacuum draws fuel and air in from the carburetor. This is called the *working fluid.* During the intake stroke, the intake valve opens as wide as possible. The goal is to take in a volume of fuel and air that is equal to the piston displacement of the cylinder. If that were possible, the engine would be operating at a volumetric efficiency of 100 percent. A normally aspirated engine will always have a volumetric efficiency of less than 100 percent, however, for several reasons. The primary reason is that a common intake manifold has numerous bends and turns built into it. These bends combined with

the intake manifold's rough metal surface restrict the flow of air through the manifold to the intake valve. Improper timing of the intake and exhaust valves is another culprit. Obviously, if the intake valve doesn't remain open long enough, the volume of working fluid will be less than normal. If the exhaust valve stays open too long or not long enough, that too, will affect volumetric efficiency. To increase the volumetric efficiency of the engine, the designers engineered the intake and exhaust valves to work together. Common sense seems to dictate that the intake valve would be wide open at top dead center and then close shut at bottom dead center, whereas the exhaust valve would open at bottom dead center and close at top dead center. That seems reasonable, but it doesn't take volumetric efficiency into consideration. Remember, the goal is to put the same volume of working fluid into the cylinder as the piston displaces. So by carefully timing when the intake and exhaust valves open and close, the engine designers have increased the efficiency. To start off, the intake valve opens at a point approximately 15 degrees before top dead center: When the intake valve is opening, the exhaust is still flowing out the exhaust valve. This creates a low pressure area within the cylinder, and the incoming fuel air mixture takes advantage of this low pressure and rushes in to fill the cylinder even as the last of the exhaust gases escape out the exhaust valve. If the intake valve opens too early, the hot exhaust gases rush into the intake manifold and ignite the incoming fuel air mixture, resulting in a backfire. (A stuck intake valve could also lead to backfiring). At approximately 15 degrees after top dead center the exhaust valve shuts and the intake valve remains open until 60 degrees after bottom dead center. The piston is actually on the compression stroke before the intake valve closes. This allows the fuel air mixture to continue flowing until the last possible second. This timing allows the maximum fuel air charge into the cylinder. As the compression stroke continues, the ignition event occurs at 28 degrees before top dead center.

After the ignition event is the power stroke. As the piston moves toward the bottom of the cylinder, the exhaust valve opens at approximately 55 degrees before bottom dead center. This helps to cool the engine by allowing the hot exhaust to begin escaping before the piston has reached bottom dead center. Most of the energy has already been used by this point; allowing the exhaust gases to escape early produces less heat buildup in the cylinder walls.

Two-Stroke Engines

Most modern aircraft engines used today are of the four-stroke variety. A small but growing segment of aviation has discovered two-stroke engines. (See Fig. 1-21.) Once relegated to hang gliders, the two-stroke engine is now appearing in numerous homemade airplanes and gyroplanes. Although the two-stroke engine is mechanically simpler than a four-stroke, it is much less efficient and harder to lubricate. Of course, its chief advantage is lower cost than a comparable four-stroke engine. The principle behind the operation of a two-stroke engine is similar to a four-stroke engine: Both engines have a five-event cycle, i.e., intake, compression, ignition, power, and exhaust. However, in the two-stroke engine, two of the five events happen almost simultaneously. This is what occurs inside a two-stroke: As the piston moves upward, a fuel air mixture is drawn into the airtight crankcase through a check valve. The piston then moves downward and compresses the fuel air mixture trapped in the crankcase. As the piston moves down, it uncovers the intake port, allowing the pressurized fuel air mixture to rush into the cylinder. The piston

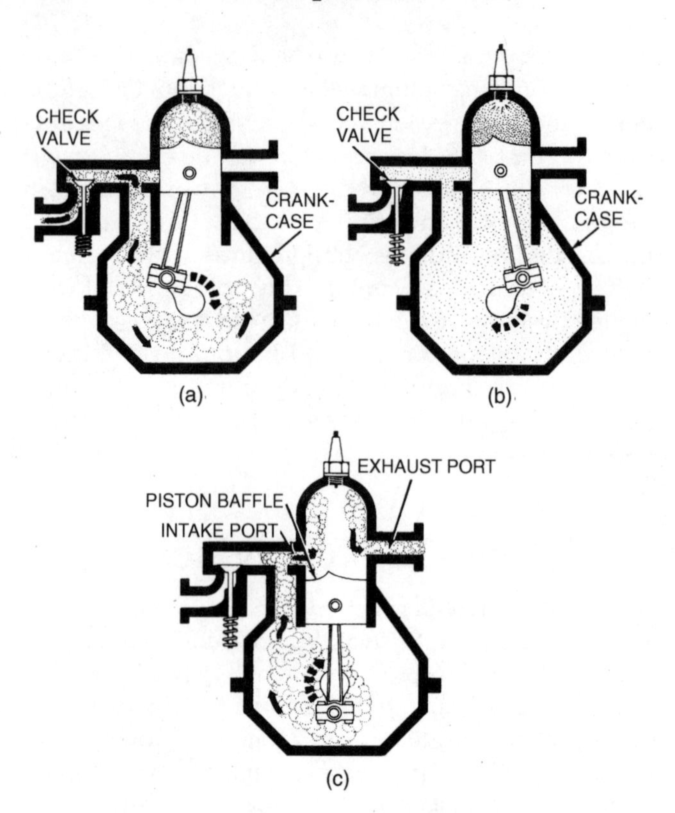

1-21 *Operation of a two-stroke engine: (a) compression event; (b) ignition and power events; (c) exhaust and intake events. (Courtesy of Michael J. Kroes and Thomas A. Wild, Aircraft Powerplants, McGraw-Hill, 1995)*

then begins to move upward on the compression stroke. Just before the piston reaches the top of the cylinder a spark is introduced which ignites the fuel air mixture. This is the power stroke. As the piston is driven downward it uncovers the exhaust port allowing the spent gases to escape. As the piston continues its downward travel, it again uncovers the intake port, starting the cycle all over again. The exhaust stroke happens almost simultaneously with the intake event. In fact, the incoming fuel air mixture mingles with the exhaust gases to some degree—the piston has a baffle built into its head to minimize this. The main difference between a two-stroke engine and four-stroke is that, for each revolution of the crankshaft of a two-stroke engine, the result is a complete cycle, whereas two complete revolutions of the crankshaft are required to complete the cycle of a four-stroke engine.

Oil

It may be a cliché, but it's true nonetheless. Oil is the lifeblood of an engine. Although lubricants can be made from animal, vegetable, or mineral origin, this section will describe mineral-based lubricants. Of the mineral-based lubricants, there are three types: (1) solid lubricants such as graphite, which are excellent for use in lubricating firearms or fishing reels; (2) the family of semisolid lubricants, which include grease, a mixture of oil and soap that has proved itself in lubricating gears and wheel bearings; (3) fluid lubricants, or oils.

Oil is a product of the distillation of petroleum. Crude petroleum is broken down into numerous products ranging from gasoline to home heating oil. Oil reduces friction. Looking at the engine parts under a microscope, one can see that the metal has numerous peaks and valleys. Metal-to-metal contact begins to wear these peaks and valleys away, causing a tremendous amount of heat and friction. To reduce the friction between moving parts, oil provides a smooth surface by filling in those peaks and valleys. Now, instead of metal sliding against metal, it's oil sliding against oil. This is where an oil's viscosity becomes very important. *Viscosity* is defined as an oil's fluid friction. In simpler terms, it means how easily an oil flows. As a general rule you want to use a lower viscosity oil in cold weather and a higher viscosity oil in hot weather, because a highly viscous oil is thick and resists pouring whereas a lower viscosity oil pours more easily. It's interesting to note that oils used in airplane engines have a higher viscosity than those used in the family car, because the aircraft engine's operating tolerances are much greater than an automobile's engine, and the oil needs to provide an unbroken film on the moving parts while the engine is operating.

In addition to reducing friction, an aircraft oil has to help cool the engine. A good oil will absorb a tremendous amount of heat from all the lubricated surfaces. This is especially important at the piston head. Not only does the oil have to absorb the heat, but it also has to dissipate it rapidly. This is done through the oil cooler. (See Fig. 1-22.) The term *oil cooler* is really a misnomer, it should be called an *oil temperature regulator.* The cooler's function is to keep the oil within a specified range during normal engine operation, primarily cooling the oil. Oil coolers come in various designs and sizes, but most incorporate a thermostatic device which allows the oil cooler to be bypassed if the oil is too cold and to route the oil through the cooler if it's hot.

1-22 *A typical oil cooler. (Courtesy of Michael J. Kroes and Thomas A. Wild, Aircraft Powerplants, McGraw-Hill, 1995)*

Another important function of the engine oil is to act as a sealant in the combustion chamber, forming a barrier between the cylinder walls and the piston rings and preventing any gases from blowing by the rings. Oil keeps the engine clean too. It carries sludge and other by-products of combustion away from the moving parts and drops them off in the oil filter. Finally, the engine oil provides a protective coating on the metal parts. This helps prevent corroding.

Mineral Oil

Mineral oil is used primarily to break in a new engine or cylinder. It contains no additives except for a small amount of polymeric chemical used to improve the oil's pour point, or ability to flow at low temperature. A problem with mineral oil is its tendency to oxidize when exposed to high temperatures.

Ashless Dispersant Oil

Almost all aircraft engine oils (other than mineral oil) contain a dispersant. The dispersant acts as a suspension agent carrying the carbon, lead, and other combustion by-products away from small oil ports or passages. If these contaminants were not suspended, they could form a sludgelike mixture that would reek havoc on small oil passages by clogging them. Or the sludge could form into deposits under the piston rings, decreasing their sealing effectiveness. The oil carries these contaminants until it reaches the oil filter or until the oil is changed. (See Fig. 1-23.) That's why an oil change is such cheap insurance.

The oil is called *ashless* because an additive is used to increase the viscosity index and provide antiwear properties. These additives leave no metallic ash behind when they burn. An engine with high ash content could have hot spots that caused preignition or spark plug fouling.

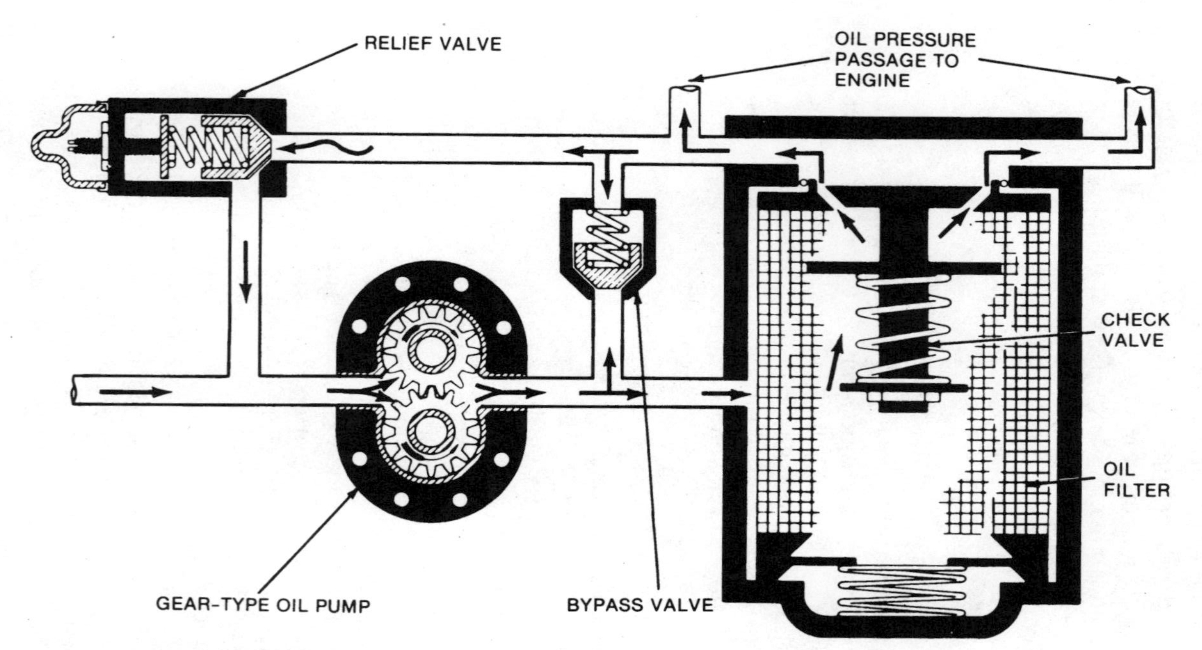

1-23 *Typical oil filtration system. (Courtesy of Michael J. Kroes and Thomas A. Wild, Aircraft Powerplants, McGraw-Hill, 1995)*

Multiviscosity Oils

Years ago, the only choice of oil was a single grade. During cold weather starts, a typical 50-weight oil takes up to 10 seconds to start to flow and protect the upper parts of the engine. Even worse, the oil can also take up to several minutes to thin out enough to be an effective lubricant. All this time the engine is wearing itself away toward an early overhaul. Although starting is the most critical time for the engine, a change in temperature can also have an adverse impact on straight-grade oil. Using a straight 30-weight oil, especially in a hot climate, may heat the oil up to the point that it starts to thin out and loose effectiveness. Using a 50-weight oil in cold temperatures may damage the engine because the oil never properly circulates. To combat these problems, oil manufactures developed multiviscosity oils. Basically, a multiviscosity oil can provide better lubrication at a wider temperature range than straight weight oil. Take a multiviscosity 10W30, for example. At cold temperature the multiviscosity oil will act like a straight 10-weight oil. As the temperature goes up, the oil takes on the characteristics of a straight 30-weight. A word of caution, if the engine is at TBO or slightly beyond, do not use a multiviscosity oil. Industry tests have shown instances of very high oil consumption in high time engines using a multiviscosity oil.

Synthetic Oil

A few years ago synthetic oil made a big splash on the marketplace. It turned out to be a belly flop. It's a great product: Synthetic oils have been used in turbine engines for years; synthetic oil has great high-temperature characteristics, resisting oxidation and the resulting deposits better than mineral oil or ashless dispersant oil; synthetic oil is a true all-weather oil—even at temperatures as low as −40°F, an engine using a synthetic oil can start and run without preheat or excess wear. But synthetic oil just doesn't have a great history in reciprocating engines. A major oil company has recently settled a case brought against them for millions of dollars in damages to aircraft engines. They agreed to overhaul or replace hundreds of engines that used their synthetic oil. If using a synthetic, check with the engine manufacturer. Not all synthetic oils are compatible with all engines (or their warranties).

Gauges

Tachometer

FAR 91.205 requires a tachometer for each engine. (See Fig. 1-24.) A tachometer is designed to provide the pilot with a means to determine an accurate reading of an engine's revolutions per minute (RPM). Depending on the age of the aircraft, the tachometer may be mechanical, magnetic, or electric. Many tachometers still in use today are of the old mechanical type. These instruments utilized a rotating shaft from the back of the engine up to the instrument itself. A coil spring and flyweight assembly in the instrument converts the rotational move-

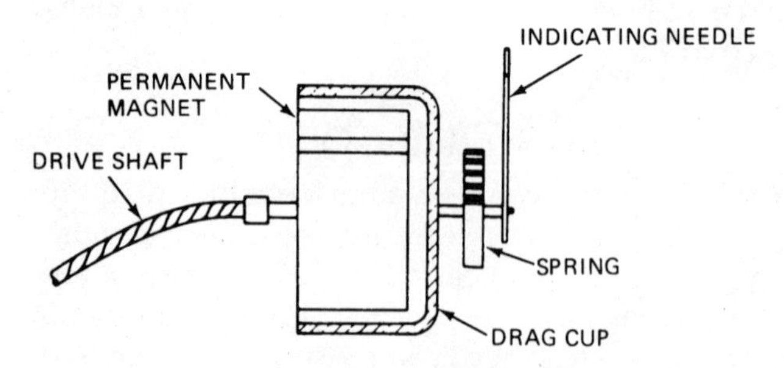

1-24 *Magnetic tachometer principle. (Courtesy of Michael J. Kroes and Thomas A. Wild, Aircraft Powerplants, McGraw-Hill, 1995)*

ment of the shaft to a reading of RPM's on the instrument's face.

The most common type of tachometer in use today is the magnetic tachometer or "tach." The magnetic tach uses a cylinder-shaped magnet which rotates in an aluminum cup called a *drag cup*. A flexible cable runs from the engine and is connected to the cylindrical magnet. As the cable turns, so does the magnet, which produces a small electrical eddy current in the aluminum cup. The eddy current produces an electromagnetic force which causes the drag cup to rotate in the same direction as the flexible cable. A spring arrangement is employed to restrict the rotation of the drag cup in proportion to the speed of the rotating magnet. Once the coiled balance spring and the drag cup's rotation are balanced, a reading of the engine RPM is possible. Because the drag cup is directly connected to the RPM needle by way of a shaft, drag cup rotation indicates engine RPM.

With magnetic tachometers, an oscillating RPM needle usually indicates a lack of lubrication. First check your tach cable for any sharp bends. Each bend in the cable should have a radius of 6 to 7 in. If the cable is free of nicks or cuts, you may want to have it lubricated with a suitable grease. This should eliminate the oscillations.

Electric tachometers are normally found on large or multiengine aircraft, although Mooney Aircraft Company uses them on some models. The electric tachometer uses a three-phase electric current in place of the flexible cable used in magnetic tachometers. A small three-phase generator is mounted on the engine and produces three-phase alternating current. The frequency of the current depends on the engine RPM. This generator is connected to a synchronous motor inside the tachometer instrument itself. The speed this motor turns depends upon the frequency of the current it receives from the generator, which in turn depends on the engine RPM. This generator is connected to a synchronous motor inside the tachometer instrument itself. The speed this motor turns depends upon the frequency of the current it receives from the generator, which in turn depends on the engine RPM. Just like its magnetic brethren, the electric tachometer utilizes a drag cup to deflect the RPM needle.

Manifold Pressure Gauge

The manifold gauge is essentially a barometer. (See Fig. 1-25.) A manifold pressure gauge is constructed with two bellows or diaphragms. One diaphragm is a sealed aneriod wafer which responds to changes in the ambient atmospheric pressure, and the other diaphragm is connected to the intake manifold of the engine. Taken together, manifold pressure is an indication of the amount of air flowing into the engine. With the throttle wide open, the engine is drawing in as much air as it can. Because of the

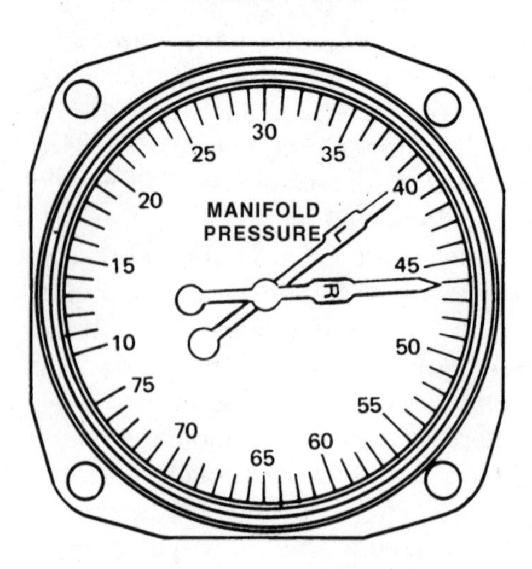

1-25 *Manifold pressure gauge. (Courtesy of Michael J. Kroes and Thomas A. Wild, Aircraft Powerplants, McGraw-Hill, 1995)*

bends and friction of moving air in the manifold itself, the pistons are trying to use more air than the manifold can supply, and hence the manifold pressure is slightly less than ambient pressure. As the aircraft climbs out, even with full throttle the manifold pressure will remain approximately 1 in below ambient pressure. Adding a turbocharger to the engine will boost manifold pressure to the point that the pistons have more air than they can use. This results in better performance of the engine. To recap, the manifold pressure gauge indicates how much air is getting to the engine, which combines air with fuel to get power.

Oil Pressure Gauge

Oil pressure gauges in light aircraft are normally a bourdon tube design. (See Fig. 1-26.) Oil pressure enters the tube at the open end of a circle-shaped tube. Since the other end of the tube is sealed, the pressure tends to make the entire tube try to straighten out. This movement is what is being indicated on the instrument. The oil pressure instrument is color-coded for easy interpretation by the pilot. As would be expected, green color bands are for normal operation and red indicate a minimum and/or maximum pressure you

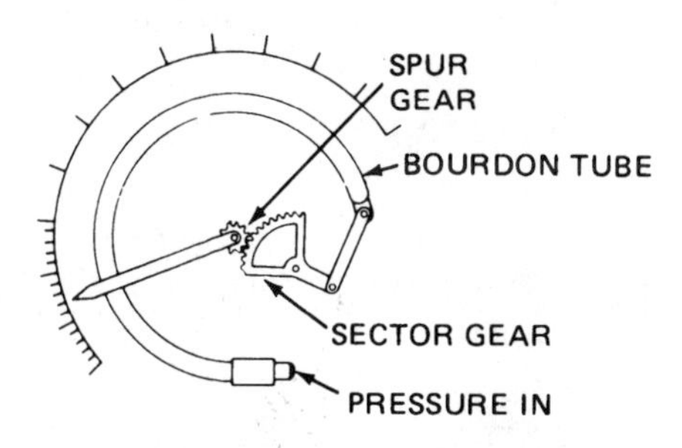

1-26 *Bourdon tube principle. (Courtesy of Michael J. Kroes and Thomas A. Wild, Aircraft Powerplants, McGraw-Hill, 1995)*

should operate at. Here's an interesting bit of trivia. Some oil pressure gauges have a black zig-zag pattern semi-imposed on the red color bands for color-blind mechanics!

Oil Temperature Gauge

Most older aircraft use a vapor pressure temperature gauge to determine oil temperature. (See Fig. 1-26.) This is a bourdon tube connected by a line to a liquid-filled bulb, submerged in the oil. As the oil heats up, it turns the volatile liquid in the bulb to a gas. This gas increases the pressure inside the bourdon tube, and it tries to straighten out. Since the pressure is proportional to the temperature, the movement of the bourdon tube accurately depicts the oil's temperature. Newer designs incorporate a temperature bulb which contains a coil of thin resistance wires. The wires from the sealed bulb lead to a circular iron core located between the poles of a magnet. Two coils are fixed on the iron core opposing each other. What they actually measure is the current flow between the coils, which is then displayed as a number or color-coded band on the oil temperature gauge.

Cylinder Head Temperature Gauge

This instrument measures the temperature of the cylinder head at the top of a cylinder. (See Fig. 1-27.) Most installations measure only one cylinder, so the accuracy of the reading is debatable. The instrument normally uses a resistance-type probe similar to the oil temperature gauge. The temperature range is usually from 240°F to around 500°F. The information is of limited value, however, since another cylinder could be melting off the engine, but if it doesn't have a probe, the pilot would never know by looking at the cylinder head temperature gauge.

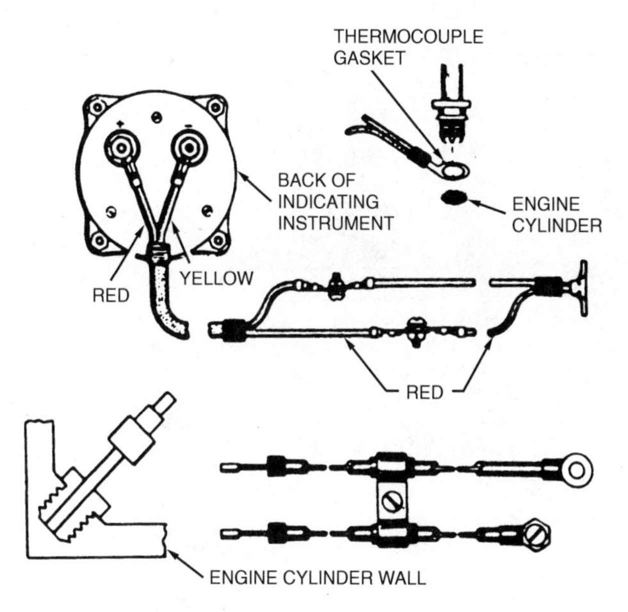

1-27 *Typical operation of cylinder head temperature gauge. (Courtesy of Michael J. Kroes and Thomas A. Wild, Aircraft Powerplants, McGraw-Hill, 1995)*

Fuel Pressure Gauges

Any fuel system which uses an engine-driven or an electrically driven fuel pump must have a fuel pressure gauge. (See Fig. 1-28.) This allows the pilot to ascertain if the system is operating properly. Normally, with a float-type carburetor, a very basic type of fuel pressure gauge will be installed. Most likely it'll have a green band between 3 to 6 lb/in^2 with red lines before and after that. This type of gauge is useful to determine if the electric fuel pump is working with the engine off or if your engine-driven fuel pump is providing adequate fuel flow to the engine with the electric pump off. If the aircraft is equipped with a fuel injection system, the fuel flow gauge becomes an indication of power output. The fuel flow gauge indicates the amount of fuel the engine is using. Some fuel flow

1-28 *Fuel pressure gauge. (Courtesy of Michael J. Kroes and Thomas A. Wild, Aircraft Powerplants, McGraw-Hill, 1995)*

gauges on injected engines are even calibrated in percentage of power. The instruments operate the same as other low-pressure gauges.

Exhaust Gas Temperature Gauge

This is probably one of the most misused and least understood instruments on the panel. (See Fig. 1-29.) An engine can run with a fuel air mixture ratio between 8:1 and 17:1. Of course, those two extremes are far from efficient. The goal is creating a combustible mixture that releases the maximum heat energy. This is accomplished by using every bit of fuel and air in the right combination, in other words, neither too rich nor too lean. By achieving this setting, the engine will be operating at peak exhaust gas temperature (EGT).

Peak EGT occurs on the EGT gauge when the exhaust gas temperature is at its maximum point. The needle peaks and starts a backward slide toward cooler temperatures. Peak EGT is the point that maximum heat energy is produced, which is not to say that

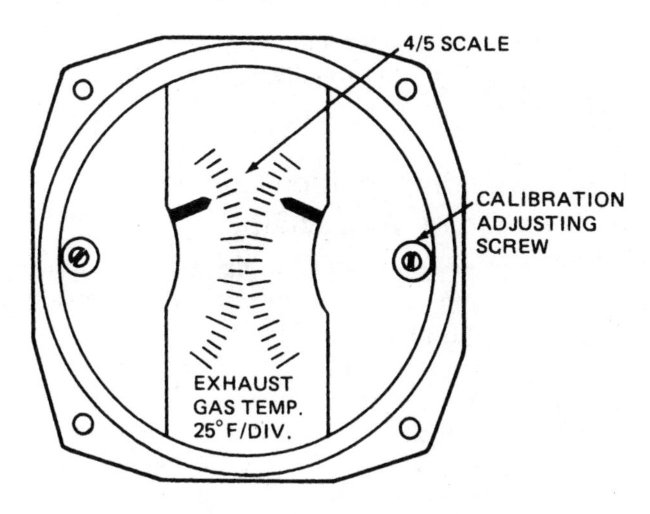

1-29 *Exhaust gas temperature gauge. (Courtesy of Michael J. Kroes and Thomas A. Wild, Aircraft Powerplants, McGraw-Hill, 1995)*

peak EGT is the best power setting. Best power is a result of pressure built up in the cylinder from unburned gases utilizing a slightly richer mixture. At approximately 100°F before peak EGT, the by-product of combustion is less dense than the completely burned mixture occurring at peak EGT. A less-dense gas takes up more volume of cylinder space and creates greater piston movement. This extra movement creates more power than the extra heat produced from operating at peak EGT. The EGT gauge is constructed utilizing a thermocouple. A *thermocouple* is a combination of two dissimilar metals that will create an electrical current which can be used to display temperature.

Maintenance

Engines are expensive to maintain and operate. Few things feel worse for a pilot than wondering if the engine is being damaged by improper leaning or poor starting techniques. Some of the best information regarding engine operations comes from the manufacturers, and the following tips are from the Lycoming Flyer.*

Avoiding Sudden Cooling of Your Engine

Sudden cooling is detrimental to the good health of the piston aircraft engine. Textron Lycoming Service Instruction 1094D recommends a maximum temperature change of 50°F per minute to avoid shock cooling of the cylinders.

Operations that tend to induce rapid engine cool down are often associated with a fast letdown and return to the field after dropping parachutists or a glider tow. On some occasions air traffic control may call for fast descents that can lead to sudden cooling.

The engine problems that may be expected when pilots consistently make fast letdowns with little or no power include:

1. Excessively worn ring grooves accompanied by broken rings
2. Cracked cylinder heads
3. Warped exhaust valves
4. Bent pushrods
5. Spark plug fouling

Generally speaking, pilots hold the key to dodging these problems. They must avoid fast letdowns with very low power (high-cruise RPM and low manifold pressure), along with rich mixtures that contribute to sudden cooling. Pilots should maintain at least 15 manifold pressure (MP) or higher and set the RPM at the lowest cruise position. This should prevent ring flutter and the problems associated with it.

Letdown speed should not exceed high cruise speed or approximately 1000 ft/min of descent.

*Used with permission.

Keeping descent and airspeed within these limits will help to prevent the sudden cooling that may result in cracked cylinder heads, warped exhaust valves, and bent pushrods.

The mixture setting also has an effect on engine cooling. To reduce spark plug fouling and keep the cylinder cooling within the recommended 50°F min limit, the mixture should be left at the lean setting used for cruise and then richened gradually during descent from altitude. The lean mixture, maintaining some power, and using a sensible airspeed should achieve the most efficient engine temperatures possible.

These operating techniques can save dollars that might be spent on maintenance. Whatever the circumstances, pilots must plan their flight operations so that the potential damage by sudden engine cooling can be avoided.

Suggestions on Engine Starts

An important part of the engine-starting procedure is the priming technique involved. Of course, the pilot's operating handbook will specify the steps in starting a specific model engine. However, some of the pilot handbooks may not explain why certain procedures are used in the starting process.

Priming can be best accomplished with an engine priming system, as opposed to use of the throttle. The primer pumps extra fuel directly into the cylinder intake port or induction system. Some float-type and pressure carburetors also provide a supplemental source of priming. Lycoming engines of more than 118 hp have a throttle pump which can be used for priming under moderate ambient temperature conditions while turning the engine with the starter.

Pilots should, however, be advised that excessive throttle priming can cause flooding of the carburetor and airbox and result in a fire in the induction system or on the outside where the fuel drains overboard. If the operator floods the engine by pumping the throttle and has a fire, it is possible to handle such a fire in the early stages by continuing to turn the engine with the starter, thereby sucking the fire back into the engine. Furthermore, if there is any fire on the outside of the engine, if the engine starts there is a good chance it will blow out the external fire.

When flooding the engine without a fire, the operator should open the throttle full and close the mixture (see operator's handbook on mixture) and turn the engine over several times with the starter to clear it. Then the pilot should begin again with a normal start routine.

Most Lycoming fuel-injected engines are simply primed by turning the fuel boost pump on, opening the mixture briefly to full rich, and cracking the throttle. Any pumping of the throttle is ineffective until the engine begins to fire.

Oil Analysis

Much is heard these days about the use of oil analysis as a tool for helping to determine engine condition. However, the vast majority of the general aviation public does not understand how this tool is to be used. We will attempt here to set forth a brief summary of the subject.

Oil analysis is not new, but it came late to general aviation as a maintenance tool. The object is to examine oil samples from an engine and break down the sample in parts per million to determine the internal health of the engine. This is based on the fact that all lubricated engine parts wear and deposit a certain amount of metallic particles in the oil. The number of particles per million of each metal determines the wear pattern for the particular engine being analyzed. It is of the utmost importance to understand that the results of the analysis *are pertinent only to the engine being analyzed,* although accumulation of data on any specific engine series is a basis for establishing standards for that series of engine.

A sharp rise above normal of the amount of a particular metal in the oil is important to note. It is imperative then to build a case history of each engine, wherein a sharp rise in any one metal will indicate abnormal engine wear. The analysis can also tell you whether the oil contains other liquid contaminants such as gasoline or water. Gasoline contamination of the oil can result from blow-by from the combustion chamber caused by poor combustion, bad timing, improper fuel mixture, worn rings, and the like. Water contamination is usually restricted to condensed vapor, but this vapor combines with the fuel combustion products to form harmful metal-attacking acids. Based on this contamination in the oil, the analysis will be able to pinpoint improper mixture, poor maintenance, and so on.

Textron Lycoming service letter L171, entitled "General Aspects of Spectrometric Oil Analysis," provides a guide for the use of oil analysis in measuring engine health. The information is in general terms, since the health of each engine must be determined on its own merits.

Differences in manufacturing processes may cause a variation in analysis results for different engine models. The amount of tin plating, copper plating, nitriding, and so on, performed during manufacture has a definite relationship to the oil analysis reports. It is not uncommon, for example, to see what seems to be high copper content early in the life of an engine, only to have this content continually decrease as the engine accumulates time and then disappears altogether.

Poor air filter maintenance, running the aircraft on the ground with carburetor/alternate air on, and holes in the air intake system are all factors that will allow an engine to ingest dirt and foreign matter. This ingestion will show up as high iron (cylinder barrels) and chrome (piston rings) content at the next oil analysis. Neither time nor space permits us here to list all the variables involved (indeed we do not profess to know them all), but it should be obvious to everyone that a continuing history of each engine is the only criteria by which its health can be determined. Remember that several samples taken at the regular oil change intervals must be analyzed to determine the normal characteristics of an engine; the first few samples on factory-fresh engines will read high as new parts are wearing in and conforming to each other.

Excessively heavy wear of internal engine parts will show up as traces in parts per million during analysis long before detrimental flaking or scoring takes place and almost always before any outward indication of trouble. This initial departure from normal is not usually any reason to tear the engine down. An investigation and timely and appropriate corrective action (replacing the air filter, perhaps) by the operator will usually result in trace elements returning to normal at the next oil change. If long TBOs are to be achieved, it is *most important* that clean air be provided to the engines.

Determining Engine Condition of High-Time Piston Powerplants

As an engine builds operating hours and approaches TBO, which may be either the manufacturer's recommended operating hours or a calendar year limit before overhaul, the question arises whether to continue flying or to do an overhaul, a major overhaul, or an exchange of engines. Here is a quick reference checklist to help make such a decision, followed by a brief explanation of the nine points:

1. Any unusual increase in oil consumption?
2. What are engine's history and calendar age?

3. How has the engine been operated?
4. What is pilot's opinion of the engine?
5. What kind of maintenance has the engine received?
6. What does the oil filter tell?
7. What has been the trend in compression checks?
8. What do the spark plugs show?
9. What are the engine manufacturer's recommendations for engine life and overhaul periods (in service letter)?

Oil Consumption

The operator and maintenance people should know what has been the general history of oil consumption during the life of an engine. A possible danger signal concerning engine health is a definite increase in oil consumption during the recent 25 to 50 hours of flight time. The oil screens and filter should be carefully observed for signs of metal. Maintenance should also take a good differential compression check at this time and should look in the cylinders with a goose neck light or a baroscope to detect any unusual conditions in the combustion chamber.

If you haven't looked at your air filter lately, carefully inspect it for wear and proper fit. This is all the more important when operating in dusty areas, and definitely could be a cause of increased oil consumption.

Engine History and Calendar Age

If a powerplant has been basically healthy throughout its life, this would be a favorable factor in continuing to operate it as the engine approached high time. If it has required frequent repairs, the engine may not achieve its expected normal life. The engine logbook should contain this accumulative record.

Another important aspect of an engine's history would be its calendar age. Engine flight time and calendar age are equally important to the operator. We have observed that engines infrequently flown do tend to age or deteriorate more quickly than those flown on a regular basis. Therefore, Textron Lycoming recommends both an operating hour limit and a calendar year limit between overhauls. Service instruction 1009 gives these recommendations, but other items in this checklist will help to determine if an overhaul or engine exchange is needed before the engine reaches these recommended limits.

Pilot's Opinion of the Engine

The pilot's opinion of the powerplant based on experience operating it is another important point in our checklist. She or he will have an opinion based on whether it has been a dependable powerplant and whether he or she has confidence in it. If the pilot lacks confidence in an engine as it approaches the manufacturer's recommended limits, this could be a weighty factor in the decision to continue flying or to overhaul it. The pilot should consult with maintenance personnel concerning their evaluation of the condition of the powerplant.

Operation

The basic question here would be how the engine has been operated the majority of its life. Some engines operating continuously at high power, or in dusty conditions, could have a reduced life. Likewise, if the pilot hasn't followed the manufacturer's recommendations on operation, engine problems and reduced life expectancy may result. This becomes a more critical influence on a decision in single-engine aircraft and for single or twin engine planes flown frequently at night or in instrument flight (IF) conditions.

Maintenance

Good maintenance should aid in achieving maximum engine life; alternately, poor maintenance tends to reduce the expected life. We notice among those powerplants coming back to the factory for remanufacture or overhaul that the smaller engines in general have had less care and attention and in a number of instances have been run until something goes wrong. The higher-power engines have generally had better maintenance and show evidence that the operators do not wait until something goes wrong but tend to observe the manufacturer's recommended operating hour or calendar limits to overhaul. The engine logbook should properly reflect the kind of maintenance provided the engine or engines. The technician who regularly cares for an engine will usually have an opinion about its health.

What Does the Oil Tell?

Clean oil has consistently been an important factor in aiding and extending engine life. A good full-flow oil filter has been a most desirable application here. When the filter is exchanged, ask the mechanic to open it and carefully examine for any foreign elements, just as is accomplished at oil change when the engine oil screen is also examined for the same purpose. Just as the spark plugs tell a story about what is going on in the engine, so the engine oil screen and the external oil filter tell a story about the health of an engine. Whether the engine is equipped with an oil filter or just a screen, oil changes should have been accomplished in accordance with the manufacturer's recommendations. These oil changes should have been recorded in the engine logbook.

If oil is analyzed, it should be done at each oil change in order to establish a baseline. Analysis is a tool which only gives useful information when a dramatic departure from the established norm occurs.

Compression Checks

What has been the trend in compression in at least the last two differential compression checks? The differential compression check is the more reliable type and should be taken on a warm engine. If the differential check reveals 25 percent loss or more, then trouble may be developing.

Some operators are confused by the compression check and its application. A compression test should be made any time faulty compression is suspected, any time the pilot observes a loss of power in flight, when high oil consumption is experienced, or when soft spots are noticed while hand pulling the prop.

Many maintenance technicians do a compression check at each oil change, and a compression check is considered part of the 100-hour engine inspection and the annual inspection. Most experienced maintenance personnel think that the differential compression check is best used to chart a trend over a period of flight hours. A gradual deterioration of charted compression taken during maintenance checks would be a sound basis for further investigation.

Spark Plugs

The spark plugs, when removed and carefully observed, tell the skilled mechanic what has been happening in the cylinders during flight and can be helpful in deciding what to do with a high-time engine:

1. Copper runout and/or lead fouling mean excessive heat.
2. Black carbon and lead bromide may indicate low temperatures, the type of fuel used, and possibly excessive richness of fuel metering at idle.
3. Oil-fouled plugs may indicate that piston rings are failing to seat or that excessive wear is taking place.
4. The normal color of a spark plug deposit is generally brownish gray.

5. In high compression and supercharged engines, a cracked spark plug porcelain will cause or has been caused by preignition.

Engine Manufacturer's Recommended Overhaul Life

Service instruction 1009 is the Textron Lycoming published recommendation for operating hour and calendar year limits until engine overhaul as they apply to each specific engine model. The amount of total operating time on an engine will be a basic factor in any decision to continue flying, change, or do a top or major overhaul of the powerplant. Operators should be reminded, however, that the hours of service life shown in the service instruction are recommendations for engines as manufactured and delivered from the factory. These hours can normally be expected provided recommended operation, periodic inspections, frequent flights, and engine maintenance have been exercised in accordance with respective engine operator's manuals.

If an operator chooses to operate an engine beyond the recommended limits, consider these factors: The cost of overhaul is likely to be greater as engine parts continue to wear, and the potential for failure may also increase.

Operators who have top-overhauled their engine at some point in the engine life invariably want to know if this extends the life of the engine. This is an important question. The chances are that if the operator applies the checklist we have been discussing and comes up with favorable answers to these questions about an engine, he or she can probably get the hours desired—with only a few exceptions. But a top overhaul does not increase the official life or TBO of the engine.

We are surprised from time to time to have owners tell us they top-overhauled their engine at some point less than the major overhaul life for no reason other than somebody said it was a good idea. Unless the manufacturer recommends it, or there is a problem requiring a top overhaul, this is a needless cost. If the engine is healthy and running satisfactorily, then leave it alone!

One other point deserves attention here; there is no substitute or cheap route to safety in the proper maintenance or correct overhaul of an engine. Apply all these basic nine points concerning your engine or engines and then make a decision whether to do a top overhaul or a major overhaul, to exchange engines, or to continue flying.

Long-Life Hints and Spark Plugs

Hot and Cold Plugs*

Today, the term *hot and cold* is commonplace in general aviation—especially when related to engine spark plugs. With the introduction of high-compression and high-horsepower engines came the need for improved spark plugs. Spark plugs used in low-compression, low-horsepower engines were not compatible with the new, more sophisticated powerplants.

Although many aircraft operators have come in direct or indirect contact with the term *hot and cold* in conversation with other pilots or mechanics, its meaning and relationship to engine operation is sometimes rather vague. What do we mean by hot and cold spark plugs? What is the relationship between an engine and spark plugs? How important is it to smooth engine operation?

Both spark plug and engine manufacturers working together determine the proper type of spark plug for each engine model. These plugs can be either fine

*Courtesy of Champion Spark Plug.

wire or massive electrode type. Before being released for production, each new type of plug is checked in the laboratory and under actual flight conditions. Plugs are tested through a wide range of operating conditions and at different power settings, and only after both engine and spark plug manufacturers are completely satisfied with test data are plugs released for production. To eliminate any possibility of error in spark plug selection, both manufacturers provide spark plug charts as a guide for proper plug selection. Final authority concerning proper plugs for a specific engine rests with the engine manufacturer.

Operating temperature of the spark plug insulator core nose is one factor that governs formation of troublesome combustion deposits. To help overcome this problem, spark plugs with the proper heat range should be selected. Spark plugs are susceptible to carbon deposits when the operating temperature of the core nose insulator is at or below 800°F, but an increase of just 100°F is sufficient to eliminate formation of these deposits. Also, lead deposits form because the bromide scavenger contained in tetraethyl lead is nonactive at low temperatures. At 900°F temperature, the bromide scavenger is fully activated, disposing of lead deposits with combustion gases during exhaust cycle. Thus an increase of just 100°F is sufficient to make the difference between a smooth and rough running engine. Operators also should avoid prolonged idling at low RPM, avoid power-off letdowns, and, after flooded starts, run engine at medium RPM before taxiing.

Deposits formed between 1000 and 1300°F are low in volume and electrical conductivity and are least apt to cause spark plug fouling. This is the reason for selecting a plug that will operate within the aforementioned temperature range at all power settings.

Normally, a hot plug is used in a cold engine with low horsepower and a cold plug in a hot engine with high horsepower. In actuality, these terms refer to the plug's ability to transfer heat from its firing end to the engine cylinder head. To avoid spark plug overheating where combustion chamber or cylinder head temperatures are relatively high, a cold plug is recommended, such as in a high-compression engine. A cold-running plug has the ability to transfer heat more readily. A hot-running plug has a much slower rate of heat transfer and is used to avoid fouling when combustion chamber and cylinder head temperatures are relatively low.

Smooth Engine Operation

Spark plugs are frequently blamed incorrectly for faulty engine operation. Replacement of old spark plugs may temporarily improve poor engine performance because of the lessened demand new spark plugs make on the ignition system, but this is not the cure-all for poor engine performance caused by worn rings or cylinders, improper fuel/air mixture, a mistimed magneto, dirty distributor block, worn ignition harness, or other engine problems. Analyzing the appearance of spark plugs that are removed from the engine may help to identify problems with the engine.

Interpreting the Appearance of Insulator Tip Deposits

The firing end of the spark plug should be inspected for color of the deposits, cracked insulator tips, and gap size. The electrodes should be inspected for signs of foreign object damage and the massive type also for copper runout.

The normal color of the deposits usually is brownish gray with some slight electrode wear. These plugs may be cleaned, regapped, and reinstalled. A new engine seat gasket should be used.

Dry, fluffy black deposits show carbon fouling. This indicates a rich fuel/air mixture, excessive ground idling, a mixture too rich at idle or cruise, or faulty carburetor adjustment. The heat range of the plug is also too cold to burn off combustion deposits.

Oil fouling will be indicated by black, wet deposits on spark plugs in the bottom position of flat, opposed cylinder engines. Oily deposits on the top plugs may indicate damaged pistons, worn or broken piston rings, worn valve guides, sticking valves, or faulty ignition supply. This same condition in a new or newly overhauled engine may simply indicate that piston rings have not yet properly seated.

Lead fouling in mild cases shows as a light tan or brown film or slight buildup on the spark plug firing end. Severe cases appear as a dark glaze, discolored tip, or as fused globules. Although mild lead deposits are always present to some degree, highly leaded fuels, poor fuel vaporization, too-cold operation of the engine, and spark plugs not suited for the particular operation are the usual causes of severe lead fouling. Extremely fouled plugs should be replaced and the cause of the fouling corrected.

Watch for bridged electrodes, a deposit of conductive material between center and ground electrodes that short out the spark plug. The gap may be bridged by ice crystals that form while trying to start, by carbon particles, by lead globules, by metallic particles, or by ingestion of silica through the air intake. When metallic fusion bridges the electrodes, the plugs must be replaced, but other deposits may simply be removed and the plugs returned to service. Deposits that short out spark plugs require corrective action.

Electrical and gas corrosion wear out spark plug electrodes. Under normal conditions, this wear occurs slowly and should be expected. Severe electrode erosion and necking of fine wire ground electrodes indicate abnormal engine operation. Fuel metering, magneto timing, and proper heat range should be checked. Spark plug cleaning and rotation at scheduled intervals is usually adequate care until spark plug gap approaches recommended maximum. Spark plugs with worn electrodes require more voltage for ignition, and should be discarded when electrodes have worn to half their original size.

Copper runout is caused by very high temperatures associated with detonation or preignition and occurs when high temperatures perforate or burn away the end of the nickel center electrode sheath and expose the copper core. Melted copper then runs onto the tip surface and forms globules or a fused mass across the electrode gap. The engine must be inspected and the plugs replaced with new ones.

A hot spot in the cylinder may cause preignition, which can always be detected by a sudden rise in cylinder head temperature or by rough engine operation. When plugs are removed after a period of preignition, they will have burned or blistered insulator tips and badly eroded electrodes.

Detonation is the sudden and violent combustion of a portion of the unburned fuel ahead of the flame front. It occurs part way through the burning cycle when the remaining unburned fuel suddenly reaches its critical temperature and ignites spontaneously. There is severe heat and pressure shock within the combustion chamber that will cause spark plugs to have broken or cracked insulator tips along with damage to the electrodes and lower insulator seal. Engine parts such as the piston, cylinder head, and connecting rod may suffer serious damage. When detonation has occurred, the cylinder must be examined with a baroscope and may require replacement. Corrective action is imperative.

The cylinders from which spark plugs with the above conditions were found should be inspected with the aid of a baroscope. Replacing the cylinder may be desirable, especially if backfiring was reported by the flight crew. If the engine operated under some detonation conditions—but not to the extent that it caused a complete piston failure—the piston rings could be broken, and a piston failure requiring a complete engine change may show up at a later date.

Damage from Excessive Temperatures

Overheating of the spark plug barrel, sometimes caused by damaged cylinder baffles or missing cooling air blast tubes, may seriously deteriorate the ignition leads. Any overheating of the spark plug barrel by a defective baffle or exhaust gas leakage at the exhaust pipe mounting flange can generate temperatures in the insulator tip sufficient to cause preignition and piston distress.

Other Spark Plug Problems

The cure for threads that are stripped, crossed, or badly nicked is replacement of the spark plug with a new one. Dirty threads in the engine may cause the spark plug to seize before it is seated. Dirty threads also cause poor contact among the spark plug, spark plug gasket, and the engine seat. This results in poor heat transfer and will cause excessive overheating of the spark plug. This condition can be corrected by making sure that threads are clean and by observing the torque specifications when installing new plugs.

Connector well flashover is caused by an electrical path along the surface of the insulator, from contact cap to shield, and occurs when the voltage required to arc across the electrode gap exceeds the voltage required to track over the surface of the insulator. This condition is caused by a too-wide electrode gap, oil, moisture, salt track, or other conductive deposit on the terminal well surface or lead-in assembly. When flashover occurs, combustion chamber residues quickly coat the insulator tip and electrodes so that the condition may be interpreted as oil or gas fouling. If the ceramic of the plug is not broken, the plug may be cleaned and reused. Thorough cleaning of the lead in assembly may solve the problem, or replacing the assembly may be necessary to affect a cure.

The size of the electrode gap has a very definite effect on spark plug service life and on the performance of the engine. Insufficient gap size will cause misfiring not only during idle but also during cruise power with lean fuel/air mixture. This intermittent misfiring during cruise lowers the temperature of the insulator tip to such an extent that lead deposits forming on the insulators may not vaporize sufficiently to keep the tips clean.

Spark Plug Servicing

The following are some hints for servicing spark plugs:

Spark plugs with cracked, broken, or loose insulators or highly worn electrodes should not be cleaned and reused. Replace with new aircraft spark plugs.

To prevent damage to spark plugs during removal and installation, use the right tools for the job.

It is helpful to use antiseize compound or plain engine oil on spark plug threads starting two full threads from the electrode, but *do not use* a graphite-based compound.

If a spark plug is dropped, discard it and replace with a new one.

A torque wrench, reading in either foot-pounds or inch-pounds, is essential to proper installation of spark plugs.

Is Your Spark Plug Connector Overtorqued?

This is a brief summary of the Champion Spark Plug instruction for connecting the spark plug to the connector: Terminal sleeves should be handled only with clean, dry hands. Before installation, wipe off the connector with a clean, lintfree cloth moistened in methylethylketone, acetone, wood alcohol, naphtha, or clean unleaded gasoline. Make certain that the inside of the spark plug shielding barrel is clean and dry. Then, without touching the connector or spring with the fingers, insert the assembly in a straight line with the spark plug. Screw the connector nut into place finger-tight, then tighten an additional ⅛ turn with the proper wrench. Damaged threads or cracked shielding barrels may result if the connector nuts are tightened excessively. Avoid excessive side load while tightening.

Rotating Spark Plugs: The Positive and the Negative

The policy of rotating spark plugs from top to bottom has been practiced by mechanics and pilots for many years. It is common knowledge in the industry that the bottom plugs are always the dirty ones and the top plugs are the clean ones. By periodically switching the plugs from top to bottom, you get a self-cleaning action from the engine whereby the dirty plug placed in the top is cleaned, and the clean plug replaced in the bottom gradually becomes dirty. Based on this cleaning action, a rotation must be established.

Because of the ever-increasing cost of aircraft maintenance and a desire to get the maximum service life from your spark plugs, the following information is offered on the effects of constant polarity and how to rotate plugs to get maximum service life. The polarity of an electrical spark, either positive or negative, and its effects on spark plug electrode erosion has long been known but had little effect on spark plug life in the relatively low-performance engines of the past. However, in the later, high-performance, normally aspirated and turbocharged engines where cylinder temperature and pressure are much higher, the adverse effects of constant polarity are becoming more prevalent. When a spark plug installed in a cylinder is fired negative and is allowed to remain there for a long period, more erosion occurs on the center electrode than on the ground electrode, and when a spark plug is fired positive, more erosion occurs on the ground electrode than on the center electrode. From this we can see that a periodic exchange of spark plugs fired positive with those fired negative will result in even wear and longer spark plug service life.

To get a polarity change, as well as switching the plugs from top to bottom, the following rotational sequence is suggested. First, when removing the spark plugs from the engine, keep them in magneto sets. After the plugs have been serviced and are ready to be reinstalled in the engine, make the following plug exchange: For six-cylinder engines, switch the plugs from the odd-number cylinders with the plugs from the even-number cylinders. For example, switch 1 with 6, 2 with 5, and 3 with 4. For four-cylinder engines, you must switch 1 with 4 and 2 with 3. During the following operating period, each plug will be fired at reverse polarity to the former operating period. This will result in even spark plug wear and longer service life. This rotational procedure works equally well on all four- and six-cylinder Lycoming engines except four-cylinder engines equipped with the single-unit dual magneto. This is a constant polarity magneto, and the only benefit to be gained by rotating the plugs is the reduction of lead deposit built up on the spark plugs when a rotation is established and followed. Another exception occurs with a few four-cylinder engines where one magneto will fire all the top spark plugs and the other magneto will fire all the bottom spark plugs. If the plugs are rotated as previously recommended, a polarity change will result, but, since the plugs do not get moved from top to bottom, no self-cleaning action by the engine will occur. Thus the bottom plugs may need cleaning at regular intervals as these are always the dirtiest. For those engines with magnetos which fire all top or bottom spark plugs, the choice of rotating plugs to change polarity or to obtain bottom-to-top cleaning action must be made by the aircraft owner or the A & P mechanic.

Maintenance Suggestions from the Lycoming Service Hangar

Spark plugs are an important engine accessory—they do such an important job so well, yet are often taken for granted. This little fellow has character. For the alert, knowledgeable mechanic, Mr. Plug is ever willing to reveal his secrets pertaining to the health of the engine's fuel system, oil consumption, combustion chamber, and even the engine treatment given by the pilot. At the Textron Lycoming Service Hangar, we have come to lean heavily on Mr. Plug's ability to "tell a story." Actually, he's our ace troubleshooter. To make it possible for Mr. Plug to do even a better job, we are listing some dos and don'ts directed at both the mechanic and pilot.

The massive electrode type spark plugs are the least expensive and do a fine job. The fine wire platinum plug is more expensive but gives longer life, is less prone to frosting over during cold starts, and appears to be less susceptible to lead fouling. The more expensive fine wire iridium plug has all the qualities of the platinum plug, and the iridium material resists lead salts erosion to a much greater degree than platinum. This results in longer plug life.

Don't reuse spark plug gaskets.

Do use the recommended torque when installing plugs.

Don't be a throttle jockey. For years we have been preaching that engines don't like sudden throttle movement. Well, the spark plugs don't like it either.

Do, after a successful flooded start, slowly apply high power to burn off harmful plug deposits.

Don't close throttle idle on any engine. Fuel contains a lead scavenging agent that is effective only when the plug nose core temperature is 9000°F or more. To attain this temperature, you need a minimum of 1200 RPM (TIG0541 is an exception). Besides, the engine's fuel system is slightly rich at closed throttle idle. This ends up giving Mr. Plug a sooty face.

Don't fly with worn or dirty air filters or holes in induction hoses and air boxes, for this is the fastest way to wear out an engine. Mr. Plug doesn't like it either. One of his worst enemies is silicon (a fancy name for dirt).

Don't attempt to clean lead deposits from plugs with an abrasive cleaner (an excellent way to keep the plug manufacturer on overtime filling replacement orders). Use the vibrator cleaner sold by the plug manufacturers. Then, sparingly use the air-powered abrasive.

Do properly lean your engine in flight as recommended by the pilot's operating handbook. In addition to being helpful to the engine in many ways, proper learning also helps the plugs run cleaner, more efficient, and longer.

Do be a little more careful in gap setting of massive electrode plugs. The top and the bottom of the ground electrode should be parallel with the center electrode.

Don't reuse obviously worn plugs. More than 50 percent of the ground electrode eroded away; the center electrode shaped like a football; the center core of the ground electrode badly dimpled? If the answer is yes, replace.

Do use antiseize compound when reinstalling plugs (caution—only sparingly on the first three threads), here is not a case of twice as much being twice as good).

Don't use any spark plug that has been dropped. One manufacturer says "If you drop it once, drop it twice—the second time in the trash barrel."

Don't reuse any plug with cracked porcelain, regardless of how it may have been working or how it bomb-checked. It will cause serious preignition.

Don't shrug off oily spark plugs. New, topped, or majored engines with some oil in the plugs is normal because rings haven't seated. A high-time engine with oily plugs means rings are wearing out. One oily plug with others dry indicates a problem in the cylinder with the oily plug. (The bottom plugs are always first to tell the story.)

Don't clean plugs with a powered wire wheel. This is known as "a slow death on a fast wheel."

Don't determine replacement spark plugs by referring to model numbers on old plugs in the engine. The mechanic ahead of you may have installed the wrong model. Use the manufacturer's chart on all plug replacements.

Although it's common practice to remove and clean the spark plugs every 100 hours, this practice may actually shorten the life of your plugs. Frequently using a thin, sharp tool to loosen the lead deposits can cause the center insulation of the spark plug to crack or break. If this occurs, discard the plug. Table 1-1 lists other common engine problems. This chart can be very helpful to pinpoint an engine problem without

Table 1-1. Engine Troubleshooting Chart

Problem	Then the problem may be	To fix the problem try
If the engine turns over but does not start	Lack of fuel	Check the fuel valves
	Engine overprimed	Clear engine of fuel, then follow the correct starting checklist
	Induction system leaks	Correct any leaks in system
	Starter slips	Replace the starter
If the engine will not idle	Propeller lever set for low RPM	Place the propeller lever in *high* RPM
	Improperly adjusted carburetor	Adjust carburetor
	Fouled spark plugs	Changing the spark plugs
	Air leak in intake manifold	Tighten the loose connection or replace damaged part
If the engine runs rough at idle	Improperly adjusted carburetor or fuel-injection system	Adjust system as required
	Fouled spark plugs	Clean or replace spark plugs
	Improperly adjusted fuel control unit	Adjust fuel controls

	Dirty or worn hydraulic lifters	Clean or replace hydraulic lifters
	Burned or warped exhaust valves or seats	Replace exhaust valves
If the engine misses at high RPM	Hydraulic tappet worn or sticking	Replace the tappet
	Warped valve	Replace the valve
	Weak breaker spring in the magneto	Repair the magneto
	Plugged fuel nozzle	Clean or replace the fuel nozzle
	Broken valve spring	Replace valve spring
If the engine runs rough at high RPM	Loose mounting bolt or damaged mount pads	Tighten or replace the mountings
	Plugged fuel nozzle	Clean the fuel nozzle
	Propeller out of balance	Remove and repair propeller
	Ignition system malfunction	Troubleshoot ignition system and repair

(*Cont.*)

Table 1-1. (*Continued*)

Problem	Then the problem may be	To fix the problem try
If the engine produces low power	Leaking exhaust system	Correct exhaust system leaks
	Ignition system malfunction	Troubleshoot ignition system and correct the malfunction
	Carburetor or fuel-injection system malfunction	Troubleshoot and correct malfunction
	Engine valves leaking or the piston rings worn or sticking	The engine may need an overhaul.
If the engine has high cylinder-head temperature	Octane rating of fuel too low	Drain fuel and fill with correct grade
	Improper manual leaning procedure	Use leaning procedure set forth in the operator's manual
	Bent or loose cylinder baffles	Inspect for condition and correct
	Dirt between cooling fins	Remove any dirt
	Exhaust system leakage	Correct any leaks
If the spark plugs continuously foul up	Worn or broken piston rings	Engine may need an overhaul
	Incorrect type of spark plugs	Install proper range spark plugs

If the engine leaks oil	Damaged seals, gaskets, or O rings	Repair or replace as necessary to correct the leaks
If the engine has low compression	Excessively worn piston rings and valves	Engine may need an overhaul
If the oil pressure fluctuates	Low oil quantity	Determine cause of the low oil quantity and then refill with oil

wasting valuable time (especially when you are paying a mechanic to troubleshoot an engine problem).

Clearly there is more to spark plugs than just buying a set and installing them in your engine. Be sure you know what type of spark plugs to use with your engine. Also, good spark plug service and maintenance is as important as proper plug selection. Care in selecting and maintaining your plugs can result in many additional hours of smooth engine operation. Additional spark plug information is always available from the engine or spark plug manufacturers and other service organizations.

Additional information about spark plugs and their servicing may be obtained by writing Champion Spark Plug Division, P.O. Box 686, Liberty, SC 29657.

Owner Maintenance

Replacing or Cleaning Spark Plugs

On the list of preventive maintenance that an owner can accomplish without the services of a mechanic is replacing or cleaning spark plugs. Be sure to utilize the appropriate service or maintenance manual for the aircraft. The following is a brief summary of the steps necessary to service aircraft spark plugs.

Tools

The following tools are required for this procedure.

1. Open-ended or box wrench of the size to fit the terminal connectors.
2. Deep socket wrench and ratchet driver. The wrench should be a 6-point wrench rather than a 12-point. A 6-point seats better on the spark plug.
3. Spark plug tray. Use the tray to keep track of the cylinder from which the spark plug was removed. If a tray is unavailable, a discarded

egg carton that has been cleaned and numbered also works well.

4. Spark plug feeler gauge. Be sure to use round wire gauge.
5. Antiseize compound.
6. Torque wrench.

Procedures

In general, spark plugs are removed from the engine in the following manner.

1. Using an open-ended wrench, remove the terminal connector from the top of the spark plug. Be careful to avoid damaging the ignition lead or the threads on the top of the plug. Pull the ignition lead straight out of the plug. (See Fig. 1-30.)
2. Use the deep socket and ratchet driver to remove the spark plug from the cylinder. Be sure the socket has a good fit on the plug before attempting to loosen it. A poorly fitting socket can damage the spark plug. (See Fig. 1-31.)

1-30 *Use an open-ended wrench to remove the terminal connector from the top of the spark plug.*

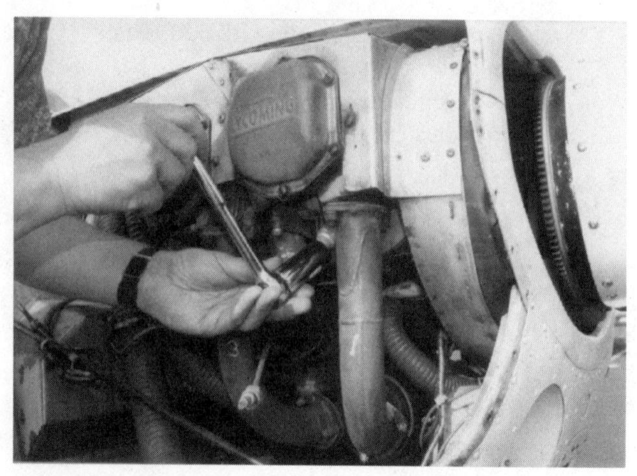

1-31 *Use a deep socket to remove the spark plug from the cylinder.*

3. Place the removed plug in the appropriately marked tray. The plug should be placed in a hole corresponding to the cylinder from which it was removed. Exercise care not to drop the spark plug: Dropping the plug may result in cracks forming in the ceramic insulation; because these imperfections may not be obvious, any dropped plug has to be rejected as unairworthy.
4. Inspect the removed plug for any excessively eroded electrodes or cracked insulation. Discard any obviously damaged or unserviceable plug.
5. Oil should be removed by use of an approved solvent. Dry the plug with compressed air, being careful to dry fully the solvent off the plug.
6. Remove any carbon deposits by utilizing an abrasive blasting machine. Most FBO maintenance shops are equipped with one. For those individuals who have limited experience with this type of machine, the best option may

be to allow the maintenance shop to clean the carbon off the plugs. Excessive blasting may wear the electrodes beyond serviceable limits.

7. Use a round wire feeler gauge to measure the gap between the electrode and the prongs. The feeler gauge will have two wires protruding from the end. The wire with the smallest diameter should pass through the gap in the spark plug. The large diameter wire should not be able to pass through the gap. If the gap is too large, it can be adjusted using a special gap-setting tool. No attempt should be made to close a too-large gap by means of pliers or other hand tools. If the gap is too narrow on a two-pronged spark plug, the best bet is to discard it. Fine wire plugs may be able to be adjusted. See the appropriate manufacturer's bulletins for further information.
8. When reinstalling the spark plugs use only new gaskets. Spark plugs wear unevenly due to the constant polarity of the magnetos used in engines with an even number of cylinders. A plug that fires with positive polarity will show excessive ground electrode wear. A plug which fires with negative polarity will show excess center electrode wear. To help equalize spark plug wear, it's important to reinstall the plugs by the following chart:

If the spark plug was removed from	*It should be replaced in*
The top of an odd-numbered cylinder	Bottom of an even-numbered cylinder
The bottom of an odd-numbered cylinder	The top of an even-numbered cylinder
The top of an even-numbered cylinder	The bottom of an odd-numbered cylinder
The bottom of an even-numbered cylinder	The top of an odd-numbered cylinder

Once the correct cylinder is located, install the spark plug and new gasket. An antiseize compound should be applied to the spark plug threads prior to inserting them. Be sure not to allow the compound to get inside the plug. Tighten as much as possible by hand. Next use a torque wrench set to the specified torque to finish tightening the plugs. As a general guide, 18-mm spark plugs are torqued between 360 and 420 in-lb, and 14-mm plugs are usually torqued to around 240 to 300 in-lb. Reconnect the ignition lead being careful not to contaminate the terminal sleeve. Any dirt or moisture that gets into the connection can short out the spark plug. Do not overtighten the terminal nut. Tighten it as far as possible by hand, then, using an open-ended wrench, apply another quarter turn. This is sufficient and will not damage the threads.

9. Make the appropriate entry into the aircraft logbooks.

Oil Change

Changing the aircraft's engine oil is another example of a relatively easy task that a noncertified mechanic can perform. (See Fig. 1-32.)

Tools

The following tools are needed for oil changes:

Open-ended wrenches of the appropriate size
Screw drivers
Oil filter wrench
Safety wire pliers and safety wire
Bucket, funnel, and short piece of hose

Procedure

Use the following steps to change the oil and filter of a typical light aircraft:

1-32 *Place the cowling in a safe spot after removal.*

1. Using the correct screw drivers, remove the engine cowling and put it in a safe place.
2. After chocking the aircraft, start and run the engine long enough to warm the engine oil.
3. After shutting down the engine, place a bucket underneath the oil drain. Insert a short piece of hose into the outlet of the funnel. Using the appropriate wrench, loosen and remove the oil drain bolt. On some installations, the safety wire will have to be removed first. Place the funnel into the draining oil stream. This will prevent oil splashing on the engine mounts or other aircraft structures. Allow the oil to drain completely.
4. Cut the safety wire and, using the oil filter wrench, remove the oil filter. (See Fig. 1-33.) Using tin snips, cut open the oil filter and inspect the paper filter element. Look for any metal or other type of debris. It may be necessary to use a solvent to clean the paper element in order to thoroughly and completely inspect it. Use a magnet to pick

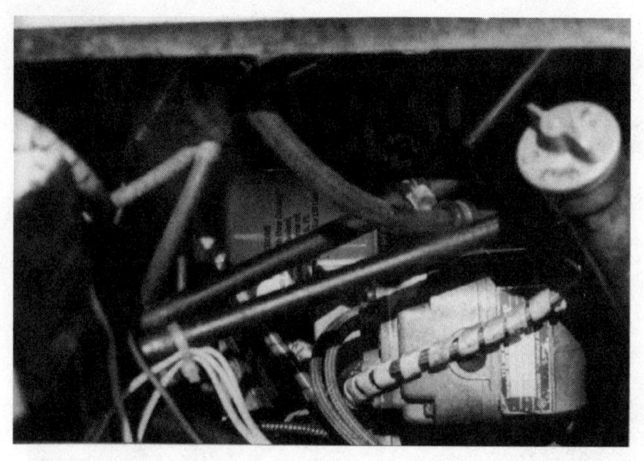

1-33 *Use an oil filter wrench to remove the oil filter.*

up any suspected metal particles. If metal is present, solicit the advice of a qualified mechanic before further flight. Some engines have an oil screen rather than a filter. If this is the case, remove the screen and clean in a solvent solution. Use compressed air to dry the screen.

5. Replace the oil filter. Be sure to use correct safety wiring techniques to secure the filter. (See Chap. 9.)
6. Replace the oil drain plug. Be sure to use a torque wrench if called for in the maintenance manual. Prior to replacing the plug, clean it with a solvent solution and replace any gaskets or O rings as necessary. Safety-wire the drain plug if required.
7. Fill the engine with the correct grade and amount of oil. Do not overfill.
8. Replace the engine cowlings.
9. Enter the oil change into the appropriate logbook. (See Chap. 8.)
10. Discard the used oil in a recycling pit.

Air Filter Replacement

Replacing the air filter should be accomplished according to the schedule determined by the engine manufacturer. It's a very easy task and requires a few hand tools, including a screwdriver. To replace an air filter, follow these steps:

1. Carefully move the propeller out of the way of the air filter.
2. Remove the screws which hold the bracket in place. Usually there are four screws. Retain the screws. (See Fig. 1-34.)
3. Remove the paper filter element from the bracket and discard.
4. Install the new element, being careful to observe the air flow direction arrow printed on the filter. Secure the filter to the bracket. (See Fig. 1-35.)
5. Install the filter bracket back into the cowling using the screws previously removed.
6. Make an appropriate logbook entry. (See Chap. 8.)

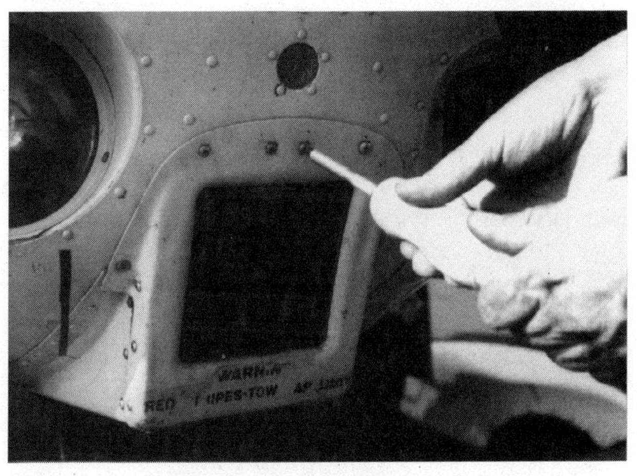

1-34 *Remove frame holding filter.*

1-35 *Example of paper element.*

Replacing an Air Cleaner

Some aircraft do not use a paper filter element. Instead, they use a foam rubber type element which can be cleaned and reused. The procedures for removing the air cleaner are essentially the same as replacing an air filter. A few simple hand tools are all that is required, including a screwdriver. To remove the air cleaner, follow these steps:

1. Using the screwdriver, remove the plastic or wire screen in front of the air cleaner. Remove the air cleaner from the air box.
2. Clean the foam air cleaner element in gasoline. Use an appropriate bucket or other type of container to avoid contaminating the element or the environment. After thoroughly cleaning the foam element, ring it out and use compressed air to dry it.
3. Rub the foam air cleaner with oil from a new can or engine oil. Allow the oil to soak throughout the foam.
4. Use compressed air to blow any excess oil from the foam element.

5. Replace the air cleaner in the air box and secure.
6. Make the appropriate logbook entry.

Fuel Line Replacement

A fuel leak or chaffing of a fuel line may be reason to change a prefabricated fuel line. Doing so requires a few hand tools, and care must be exercised while working around gasoline. Be sure to ground the aircraft and use plastic containers to catch any leaking fuel while removing the old fuel line. (See Fig. 1-36.)

Tools.

The following tools will be needed to replace a fuel line:

1. Open-ended wrench of the appropriate size
2. Screwdriver
3. Safety wire and pliers

Procedures.

To replace a fuel line, follow these steps:

1. Turn off the fuel shut-off valve. Remove the cowling to gain access to the damaged fuel line.
2. Using an open-ended wrench, loosen the connection between the fuel line and the connection point. Use care in removing the line, as residual fuel may spill out. Be sure a plastic container is underneath the work area.
3. Inspect the new fuel line for any obvious damage. Compare it to the fuel line that has

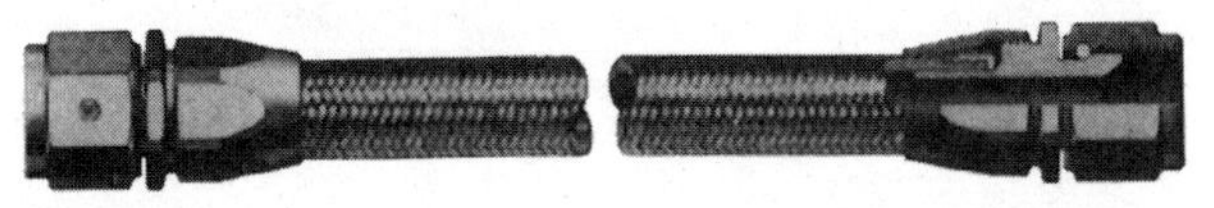

1-36 *Typical fuel line.*

been removed for correct size and shape. Install the fuel line and tighten the connections as called for in the maintenance manual and safety wire. Be sure that the fuel line does not rest on any other surfaces and does not chafe against any control lines or wires.

4. Be sure the aircraft is securely chocked and turn the fuel shut-off valve back to on. Run the engine for several minutes, then shut down and check for any leaks.
5. Replace cowling if no leaks were noted.
6. Complete logbook entry. (See Chap. 8.)

2

Airframe Repair

Wood, Fabric, and Metal

Fuselages

Most modern aircraft are of semimonocoque construction. (See Fig. 2-1.) The skin of a semimonocoque fuselage carries a large part of the load and stress imposed on the aircraft during flight. An internal framework of stringers, frames, and bulkheads also share in the load-carrying burden. A full monocoque fuselage, in contrast, does carry all the loads imposed by flight; this fuselage is basically a tube without any internal framework at all. Usually, this type of fuselage has a small diameter so that it can effectively carry the loads.

Another type of fuselage is the truss fuselage. A truss is made up of tubes and bars connected to four longitudinal members which run the length of the fuselage. How the lateral bracing is arranged will determine whether it's a Pratt truss or a Warren truss. (See Fig. 2-2.) A Pratt truss has vertical and lateral

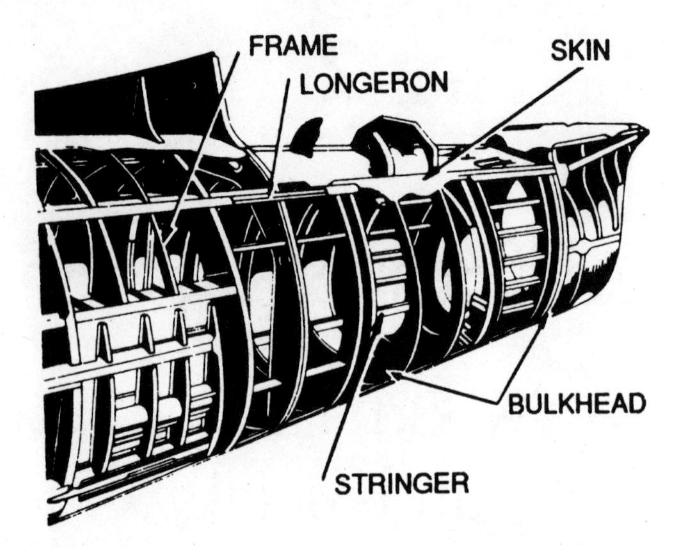

2-1 *Semimonocoque fuselage. (Courtesy of Michael J. Kroes, William A. Watkins, and Frank Delp, Aircraft Maintenance and Repair, McGraw-Hill, 1993)*

members, forming a box shape when connected to the longitudinal members. The diagonal members are usually constructed of wire and carry only tension loads. A Warren truss has diagonal tubes between the longitudinal members. They can carry tension and/or compression loads depending upon which way the load is acting. Although most truss construction is done with steel, it can also be wood, aluminum, or other materials.

Semimonocoque fuselages have the skin attached by rivets or bonding to an interior of frames and bulkheads. The thickness of the skin is determined by what location and what stresses are involved in the area the skin will be used. Table 2-1 shows a chart of typical aircraft materials and thickness. The four most common airframe building materials are wood, fabric, aluminum, and composites (not necessarily in that order).

Wood

Wood has been used to construct airplanes from the days of the Wright brothers, and it is still in use today. Although wood is now primarily employed by homebuilders, some production airplanes, such as the Bellanca Viking, are still made of wood. (See Fig. 2-3.) When selecting wood for a homebuilt project or if purchasing an aircraft utilizing wood construction, one should understand some characteristics about this product. Although many types of wood have been used, research and user experience has shown Sitka spruce to be the best overall choice for aircraft use. Sitka spruce is strong, lightweight, and stiffer than other choices. Of course, other choices are available depending on specific construction requirements. A builder or manufacturer's choice of wood is

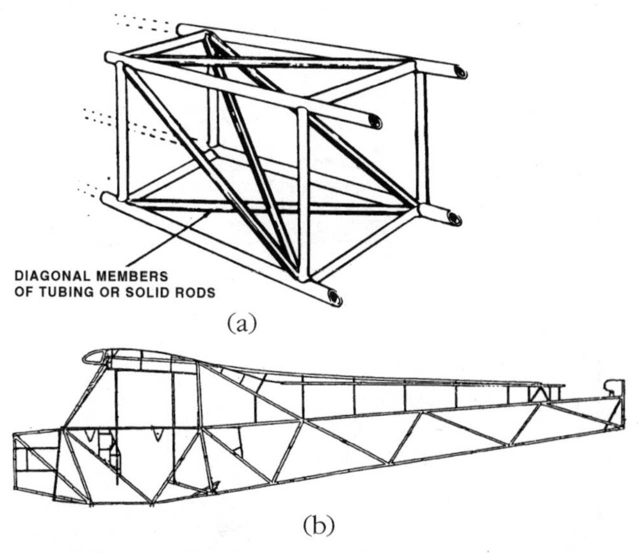

2-2 *(a) Example of a Pratt truss. (Courtesy of Michael J. Kroes, William A. Watkins, and Frank Delp, Aircraft Maintenance and Repair, McGraw-Hill, 1993)*
(b) Example of a Warren truss. (Courtesy of Piper Aircraft, Inc.)

Table 2-1 Aircraft Woods

Species	Strength properties*	Grain slope †	Remarks
Spruce (*Picea*) Sitka (*P sitchensis*), Red (*P. rubra*), White (*P. glauca*)	100%	1:15	Excellent for all users. Considered as standard for this table.
Douglas-fir (*Pseudotsuga taxifolia*)	Exceeds spruce	1:15	May be used as substitute for spruce in same sizes or in slightly reduced sizes providing reductions are substantiated. Difficult to work with hand tools. Some tendency to split and splinter during fabrication, and considerably more care in manufacture is necessary. Large solid pieces should be avoided due to inspection difficulties. Gluing satisfactory.

Noble fir (*Abies nobiles*)	Slightly exceeds spruce except 8% deficient in shear	1:15	Satisfactory characteristics with respect to workability, warping, and splitting. May be used as direct substitute for spruce in same sizes providing shear does not become critical. Hardness somewhat less than spruce. Gluing satisfactory.
Western hemlock (*Tsuga heterophylla*)	Slightly exceeds spruce	1:15	Less uniform in texture than spruce. May be used as direct substitute for spruce. Upland growth superior to lowland growth. Gluing satisfactory.
Pine, Northern white (*Pinus strobus*)	Properties between 85 and 96% those of spruce	1:15	Excellent working qualities and uniform in properties but somewhat low in hardness and shock-resisting capacity. Cannot be used as substitute for spruce without increase in sizes to compensate for lesser strength. Gluing satisfactory.

(Cont.)

Table 2-1 *(Continued)*

Species	Strength properties*	Grain slope †	Remarks
White cedar, Port Orford (*Charaecyparis lawsoniana*)	Exceeds spruce	1:15	May be used as substitute for spruce in same sizes or in slightly reduced sizes providing reductions are substantiated. Easy to work with hand tools. Gluing difficult, but satisfactory joints can be obtained if suitable precautions are taken.
Poplar, yellow (*Liriodendrow tulipifera*)	Slightly less than spruce except in compression (crushing) and shear	1:15	Excellent working qualities. Should not be used as a direct substitute for spruce without carefully accounting for slightly reduced strength properties. Somewhat low in shock-resisting capacity. Gluing satisfactory.

SOURCE: Michael J. Kroes, William A. Watkins, and Frank Delp, *Aircraft Maintenance and Repair,* New York: McGraw-Hill, 1993.

* As compared to spruce.

† Maximum permissible grain deviation.

2-3 *Bellenca Viking. (Courtesy of Michael J. Kroes, William A. Watkins, and Frank Delp, Aircraft Maintenance and Repair, McGraw-Hill, 1993)*

fairly simple, it's going to be either *hardwood* or *softwood*; these terms have nothing to do with how hard or soft a particular type of wood is but rather relate to its cellular makeup.

Some examples of softwoods used in aircraft construction are the Douglas-fir, white cedar, and Sitka spruce. As mentioned, the Sitka spruce is used as the standard-bearer with which to compare the other woods' suitability for aircraft use. Softwood trees can be identified by their needlelike leaves; the wood itself has a high strength-to-weight ratio. This makes softwood an excellent aircraft construction material. The cells in softwood are small and fibrous, and thus the wood is ideal for wing spars and other compression structures.

Hardwood trees are deciduous, that is, they lose their broad leaves each fall season. Some common examples include mahogany, white ash, and birch. The cells in these woods are a mixture of both large and small, with the large cells causing pores in the wood. The small cells are more fibrous. Usually hardwood is heavier than softwood and is used where strength more than light weight is needed. Hardwood is also used as the core material in aircraft plywood.

How to Choose Wood for Aircraft Use

The main concern when choosing wood for construction or repair is the strength of the wood itself. Is it strong enough and of sufficient quality to do the job? Another consideration is how much water is normally retained in the wood? Ideally, a moisture content between 8 and 12 percent is preferred. Anything outside of this range should be considered unacceptable for aircraft use. Kiln drying is a process designed to remove excess water from wood. The water in wood is classified as free or cell water. Free water is used by the tree to carry nutrients up and down the trunk. Cell water is the natural moisture trapped in the cellular structures of the wood. Kiln-dried wood is heated in a precisely controlled temperature oven which removes all the free water and a portion of the cell water from the wood. Kiln drying also eliminates any insect infestation. Once dried, the specific gravity should be between 0.34 and 0.40 depending on the type of wood.

The process used to cut the wood also determines its strength. Edge is the preferred method of cutting: This is a cut made so that the annual rings run parallel to the narrowest side of the board; once cut, it's important to determine the slope of the grain line. Aircraft wood is allowed a maximum slope of 1:15. Starting at the edge of the board, the grain line should run parallel for the length of the board. Since this rarely happens, a slope angle is tolerated. Again, starting at the edge of a board, the grain can move 1 in up or down for every 15 in of board length. (See Fig. 2-4.) Aircraft wood must also have a minimum number of annual rings per inch. This is known as the *grain count,* and most softwoods must have at least 6 rings per inch.

Numerous defects make wood unacceptable for aircraft use. Later in this chapter is a list of the most common wood defects. As a general rule, if there are

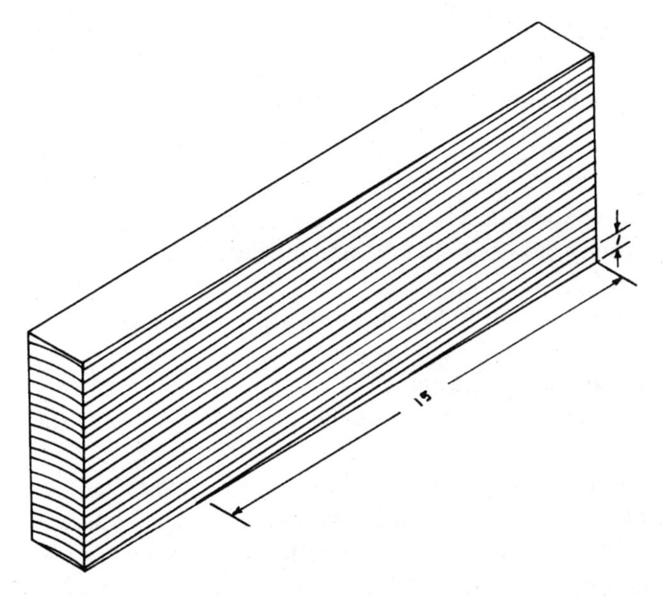

2-4 *Maximum slope allowed in aircraft wood is 1:15. (Courtesy of Michael J. Kroes, William A. Watkins, and Frank Delp, Aircraft Maintenance and Repair, McGraw-Hill, 1993)*

any doubts about the wood, don't use it. Make sure that the wood selected is defect-free and is guaranteed by the manufacturer as being aircraft quality.

Plywood

Make no mistake, aircraft plywood is stronger and of better quality than anything found at the local lumberyard. It is the only choice for aircraft use. Aircraft plywood is manufactured by gluing a number of piles or layers of mahogany and birch together in a large heated hydraulic press. Each layer of this wood veneer is assembled with the grain angle at 45 to 90 degrees to the layer beneath it. This inner layer is called the *core,* and the outside layer is the *face.* Plywood has a number of advantages over solid woods. It resists cracking better, doesn't warp very easily, and is equally strong in either width or length when subject to stress. Plywood also resists shrinkage or bowing when the moisture content changes.

Woodworking

The gluing of wood structures or members together, when done correctly, results in a structure that is as strong as the individual piece of wood. To achieve this bond, the following steps should be applied.

Preparing the Surface

Preparing the surface of the wood for gluing is the most important step in creating a strong bond. Be certain that the two pieces of wood to be joined are free of any contaminants, loose grain, or surface blemishes. If gluing softwood, do not sand the surfaces prior to gluing. Sanding will fill up the wood pores with the residue left over from the sanding process and result in a weaker bond. Sanding hardwoods, however, is an acceptable preparation method. Planing is considered the most acceptable method of surface preparation prior to gluing. Just remember to complete the gluing process within 8 hours of preparing the wood. Waiting any longer will require repeating the preparation process. The surfaces being glued must be free of paint, oil, or any type of solvents. Wax should also be removed and is usually only found in wood products, like plywood, that have been stored or otherwise protected with a waxed paper. Determine the moisture content of the wood to be glued. Look for a moisture content between 8 and 12 percent; a moisture content closer to 12 percent is more desirable. Gluing wood that is too dry may produce a starved joint that is too weak for aviation applications.

Mixing Glue

Correctly mixing glue depends on the ambient temperature. Try to work in a room with the temperature controlled at approximately 70°F (21°C). When mixing the powder or liquid glue with water, stir the two in a slow deliberate fashion. This will preclude getting any air into the mixture: Air bubbles can result in weak joints or actually cause the joint to fail. Remember the glue has a working life of 4 to 5 hours at 70°F, so have everything ready to go while mixing the glue components together. When working in an area where controlling the temperature isn't possible and the temperature is warmer than 70°F, extend the working life of the glue by immersing the glue container in a pan of cool water. Remember to always follow the manufacturer's recommendations when working with glue products.

Gluing Procedures

There are two ways to glue wood pieces together. One is called the *open assembly* method, and the other is called the *closed assembly method.* A strong bond requires a continuous layer of glue between the two pieces of wood, so try to avoid air bubbles or any kind of debris or dirt between the two pieces being joined. Spread the glue evenly over the two pieces of wood with a brush, being careful not to leave any brush hairs in the glue (this is where an expensive brush may be worth the few extra dollars). If using the open assembly method, let the two pieces stand apart for the amount of time recommended by the glue manufacturer. This will allow the glue to set up and become thicker before you join the two wood pieces together. The closed assembly method is accomplished by placing the two pieces together as soon as possible after applying the glue. Once joined, apply sufficient pressure to form a good tight seal and bond. This helps spread a thin film of glue across the entire bonded surface and will force any trapped air out from between the two pieces of wood. As a general rule, use light pressure when

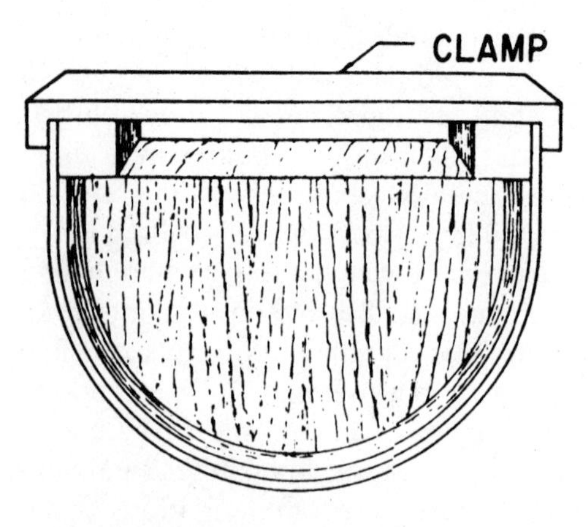

2-5 *Example of a wood clamp. (Courtesy of Michael J. Kroes, William A. Watkins, and Frank Delp, Aircraft Maintenance and Repair, McGraw-Hill, 1993)*

using a thin glue and heavier pressure when using a thick glue. Use of hands is acceptable for short durations, but wood clamps, weights, or screws are much more effective. (See Fig. 2-5.) Try to achieve a pressure range of 120 to 155 lb/in^2 for softwoods and between 150 to 200 lb/in^2 for hardwoods.

Wing Spars and Ribs

Wooden wing spars are constructed in two basic designs. They have either a solid spar or a built-up spar. (See Fig. 2-6.) A solid spar is one piece of wood or several pieces of wood laminated together. A built-up spar utilizes a combination of solid woods and plywood. These spars are designated as I beams, C beams, or box beams depending on their shape. Depending on the severity of damage, a wooden spar that has suffered an accident or some other mishap may be repaired if the costs of repair don't exceed the value of the airplane. Spars that have previous damage, damage to the wing fittings, or wood rot are

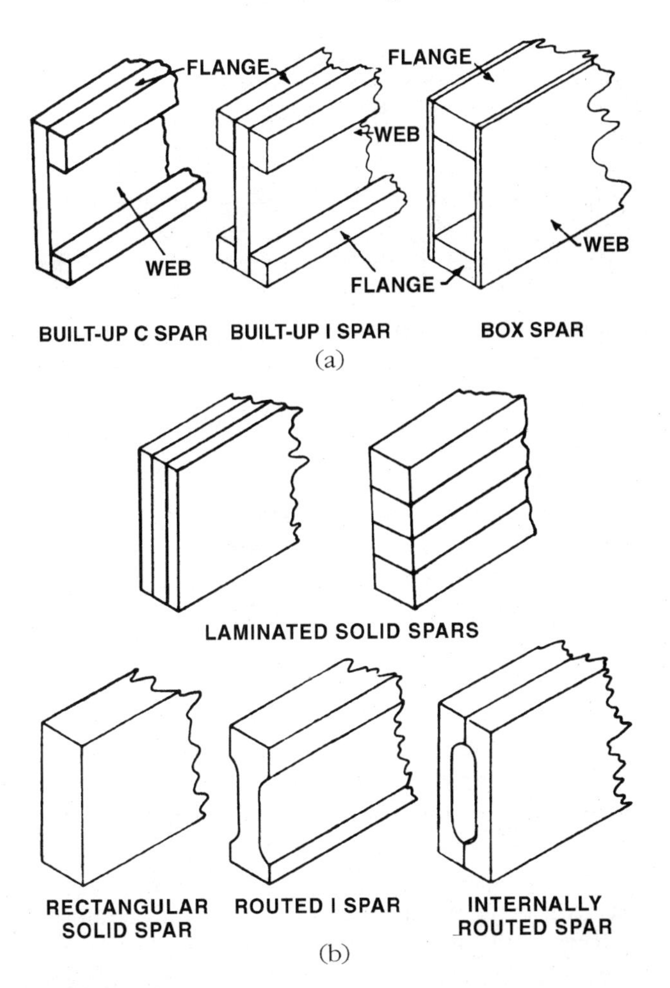

2-6 *Examples of wing spars. (a) Built-up spars. (b) Laminated and solid spars. (Courtesy of Michael J. Kroes, William A. Watkins, and Frank Delp, Aircraft Maintenance and Repair, McGraw-Hill, 1993)*

generally not repairable. Any time the strength of the spar is in doubt, consider the spar unrepairable.

Ribs give the wing its aerodynamic shape. Use the manufacturer's drawings when making or repairing a wooden rib. (See Fig. 2-7.) If, for some reason

2-7 *Rib. (Courtesy of Michael J. Kroes, William A. Watkins, and Frank Delp, Aircraft Maintenance and Repair, McGraw-Hill, 1993)*

the drawings aren't available, use an original rib as a template to create a new one.

Care of Wood

The biggest enemy to wooden aircraft is moisture; wood structures will last almost indefinitely if they are kept dry. That doesn't mean wooden aircraft can't be flown in the rain, it just means that the wood shouldn't absorb any more moisture than its natural moisture content, and use of a good spar and wood varnish on all the wood structures is imperative. Polyurethane is becoming the protectant of choice among homebuilders. Remember to clean out any drain holes to allow built-up water to escape. Wood that sits in water tends to absorb the water and swell. As the water evaporates, the wood contracts. This constant swelling and contracting will cause the wood to crack and will eventually weaken the wooden structure. If recovering the aircraft, make certain to mark the drain holes and reinstall them. Forgetting to replace just one can cause extensive damage. Temperature changes can also affect the wooden structures. Extreme heat tends to loosen glue and can cause wood to shrink. Conversely extreme cold can cause damage to wet wood by freezing the water in the cells, damaging the strength of the wood.

Diagnostic

When inspecting a wooden structure or an aircraft for airworthiness, look for the following items.

Dry Rot and Decay

Fungus can grow on damp or wet wood and, if left unattended, can break down the cellular structure of the wood. Follow these guidelines regarding dry rot and decay:

1. Always inspect for evidence of mildew. Mildew is formed by excessive humidity and heat. Extensive mildew will attack the cellulose in the wood and may develop into dry rot.
2. Make sure to inspect glue joints for any indications of loss of adhesion. Loss of adhesion is easily discernible by
   - Opening up of a joint or sign of cracks in the joint.
   - Discoloration of a glue joint. This may be caused by an adverse chemical reaction in the glue joint.
3. Inspect for loosening of any nails in a scarf patch. This could be evidence of adverse movement developing in the spar.
4. Inspect all fittings for evidence of wood shrinkage, especially at the point of installation. Wood shrinkage can be detected by loose bolts or screws.
5. It's very important to inspect for evidence of wood being excessively wet. Look for signs that the wood has been immersed by entrapped water: Fittings that have become embedded in the wood instead of flush is a sure sign of water damage.
6. Inspect for evidence of cracking. This could be caused by excessive stress, or excessively tight fittings. Look for impact damage especially if the aircraft has previous damage history.
7. Inspect for the condition of the finish. Any loss of finish could be from abrasion or

deterioration of the chemical stability of the finish. A good finish is important to help prevent attack by mildew, fungus, or oxidation.

8. Inspect for fungus. This is usually caused by "hot house" conditions, that is, moisture, heat, and exposure to spores released by fungus in the storage areas.
9. Inspect for dry rot. Wood structures which have been exposed to mildew or fungus or have excessive shrinkage, discoloration, and cracks may develop dry rot. Dry rot is easy to detect, since the wood crumbles easily under pressure.
10. Inspect for excessive wood defects. The list below will help you determine what defects are acceptable.

Permitted Defects

The following defects in wood structures are acceptable:

1. *Cross grain.* Spiral grain, diagonal grain, or a combination of the two is acceptable providing the grain does not diverge from the longitudinal axis of the material. A check of all four faces of the board is necessary to determine the amount of divergence.
2. *Wavy, curly, and interlocked grain.* These grains are usually acceptable if the irregularities do not exceed limitations specified for spiral and diagonal grain.
3. *Hard knots.* Sound hard knots of up to 3⁄8 in in maximum diameter are acceptable providing: (a) they are not in projecting portions of I beams, along the edges of rectangular or beveled unrouted beams, or along the edges of flanges of box beams

(except in lowly stressed portions); (b) they do not cause grain divergence at the edges of the board or in the flanges of a beam; and (c) they are in the center third of the beam and are not closer than 20 in to another knot or other defect (pertains to 3/8-in knots—smaller knots may be proportionately closer). Wood with knots greater than 1/4 in should be used with caution.

4. *Pin knot clusters.* Small clusters are acceptable providing they produce only a small effect on grain direction.
5. *Pitch pockets.* Pitch pockets are normally acceptable in the center portion of a beam, providing they are at least 14 in apart when they lie in the same growth ring and do not exceed 1½ in length by 1/8 in width by 1/8 in depth. Pitch pockets should not be along the projecting portions of I beams, along the edges of rectangular or beveled unrouted beams, or along the edges of the flanges of box beams.

Defects Not Permitted

The following defects in wood structure are not acceptable:

1. *Compression wood.* This defect is very detrimental to the strength of the wood and is very difficult to recognize. It is characterized by high specific gravity and has the appearance of an excessive growth of summer wood. Most species show little contrast in color between spring wood and summer wood. In doubtful cases reject the material, or subject a sample to a toughness machine test to establish the quality of the wood. Reject all material containing compression wood.

2. *Compression failures.* This defect is caused from the wood being overstressed in compression caused by natural forces during the growth of the tree, by trees falling on rough or irregular ground, and rough handling of logs or lumber. Compression failures are characterized by a buckling of the fibers. They appear as streaks on the surface of the piece at substantially right angles to the grain, varying from pronounced failures to very fine hairlines that require close inspection to detect. Always reject wood containing obvious failures. In doubtful cases reject the wood. It is permissible to make a further inspection in the form of microscopic examination or toughness test.
3. *Decay.* Examine all stains and discolorations carefully to determine whether they are harmless or in a stage of preliminary or advanced decay. All pieces must be free from rot and all other forms of decay.

Fabric

Numerous homebuilts and even some new production airplanes are covered with fabric. Fabric coverings have improved substantially since the early aviation pioneers covered their airframes with linen or silk. Today the aircraft owner has a choice between cotton, polyester fiber, glass fiber, or linen. Examine each fabric type to determine which one is most suitable for a recovering project. (See Fig. 2-8.)

The following are common terms used regarding fabrics:

Bias. A cut or fold that is made diagonally to the threads that run the length of the fabric

Calendaring. A process which produces a smooth fabric finish by ironing the nap

Fill. Fibers that are woven into a piece of fabric and run perpendicular to the length of the fabric. (See Fig. 2-8.)

Nap. Individual fibers in a fabric which give it a fuzzy texture

Mercerized. Washed in a caustic solution to remove the natural waxes

Selvage. The natural edge of a piece of fabric

Sizing. A glue used to add stiffness to the fabric; it also protects the fibers

Thread count. The number of threads per inch in a fabric; the count can be either warp or fill

Warp. Fibers that are woven into a piece of fabric and run parallel to the length of the fabric as it comes off a roll

Weight. The weight of square yard of fabric

Cotton

Grade A cotton has been used for years as an airplane cover. New synthetic fibers have made cotton almost obsolete, but some people still prefer cotton

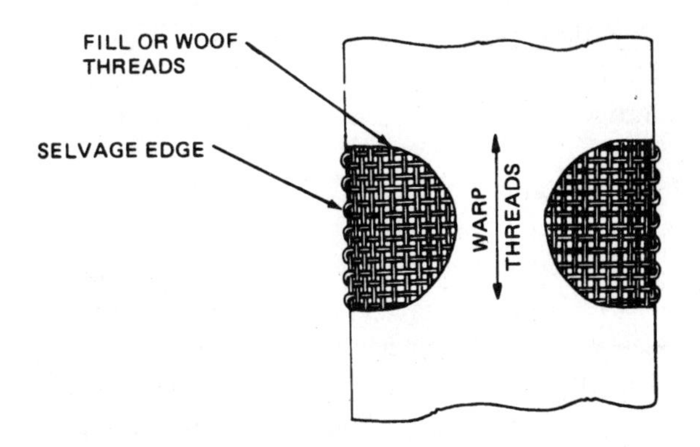

2-8 *Example of aircraft fabric. (Courtesy of Michael J. Kroes, William A. Watkins, and Frank Delp, Aircraft Maintenance and Repair, McGraw-Hill, 1993)*

to recover their machines. Most cotton fabrics weigh about 4.5 oz per square yard and have between 80 to 85 threads per inch in the warp and fill direction. Grade A cotton is mercerized but not preshrunk.

Linen

The correct name is Irish linen. This fabric is woven from flax fiber, is very strong, and will last much longer than cotton. It too, has largely passed from the scene and is normally used only in aircraft restoration projects. It's more difficult to work than cotton and weighs a bit more.

Synthetic

Polyester and fiberglass fabric have become the products of choice for use on cloth-covered aircraft. These fabrics are created by a chemical process, then woven into fabric. The polyester fabrics come in various weights and thread counts and are marketed under the names of Stits Poly-Fiber and Ceconite. The most common fiberglass product is called *Razorback.* Fiberglass coverings are virtually immune to deterioration resulting from heat and chemicals and therefore are widely used to cover agricultural aircraft. Both fiberglass and polyester fabrics resist damage from ultraviolet light and microorganisms.

Reinforcing Tape

This tape product is made out of cotton and is laid across the ribs or any other part of the aircraft structure that the fabric touches. This ensures protection of the fabric at the point of contact between the fabric and the aircraft structure. The tape comes in several sizes with ¼-, ⅜-, and ½-in widths being the most common.

Surface Tape

Surface tape is made out of the same material that is used to cover the entire aircraft. The main purpose of the tape is to protect the rib stitches and other locations where the edges of fabric meet. Grade A tape comes in 1- to 4-in widths, and the edges are notch-cut or "pinked" to prevent them from unraveling in flight. Polyester fabric also comes in 1- to 4-in widths, and the edges are heat-treated to form a permanent bond and prevent them from unraveling.

Grommets

Grommets are small plastic or brass reinforcement rings. (See Fig. 2-9.) They are used as drain holes or as lacing eyes to allow fabric to be laced together without ripping. Most plastic grommets are used for drainage purposes. Because the inside of an aircraft's structure is constantly exposed to temperature changes, grommets allow the moisture to escape. Grommets are located on the bottom and the trailing edge of horizontal flight surfaces and at the bottom of the vertical flight controls (Fig. 2-10). Additional grommets should be placed at the lowest points of the fuselage. When placing grommets along the horizontal surfaces, remember to locate them where they will best drain water. On a positive dihedral wing, for example, place the grommets along the outside of the ribs, thus keeping water from becoming trapped and damaging the rib structures. Of course, on a negative dihedral, put

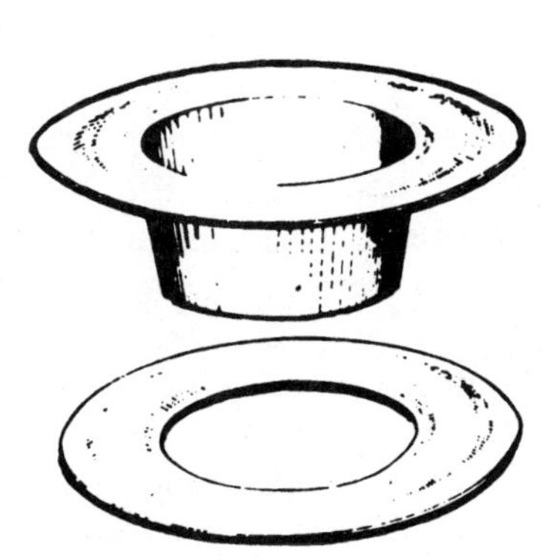

2-9
Example of aircraft grommet. (Courtesy of Michael J. Kroes, William A. Watkins, and Frank Delp, Aircraft Maintenance and Repair, McGraw-Hill, 1993)

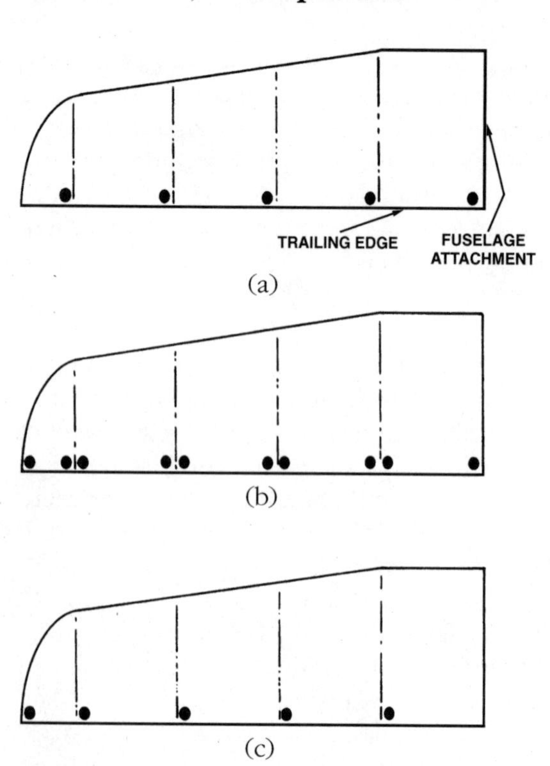

2-10 *Correct placement of grommets. (a) Positive dihedral. (b) Neutral dihedral. (c) Negative dihedral. (Courtesy of Michael J. Kroes, William A. Watkins, and Frank Delp, Aircraft Maintenance and Repair, McGraw-Hill, 1993)*

the drain grommets on the inside of the ribs. A neutral dihedral calls for grommets on both sides of the ribs. During construction of a seaplane, be sure to use seaplane grommets. During installation, after waiting until the first layer of dope (see below) has dried, simply lay the plastic grommet in the position called for by the manufacturer. Finish by doping the piece in place. Seaplane grommets have bigger openings and drain more water than regular grommets.

Inspection Ports

Installation of inspection rings (Fig. 2-11) is accomplished in the same manner as installing grommets. First, locate the exact position required by the manufacturer. If the specifications are no longer available for the airplane, use the currently installed inspection ports as a guide. Next, after the first coat of dope has dried, dope the inspection ring into position. An inspection ring is usually about 4 in in diameter, so make sure enough room is available to mount it in its proper location. Once it's in place, most builders or owners do not cut out the fabric center portion until the first annual inspection. Use a sharp razor knife, being careful not to cut or nick the ring itself. Once removed, cover the hole with

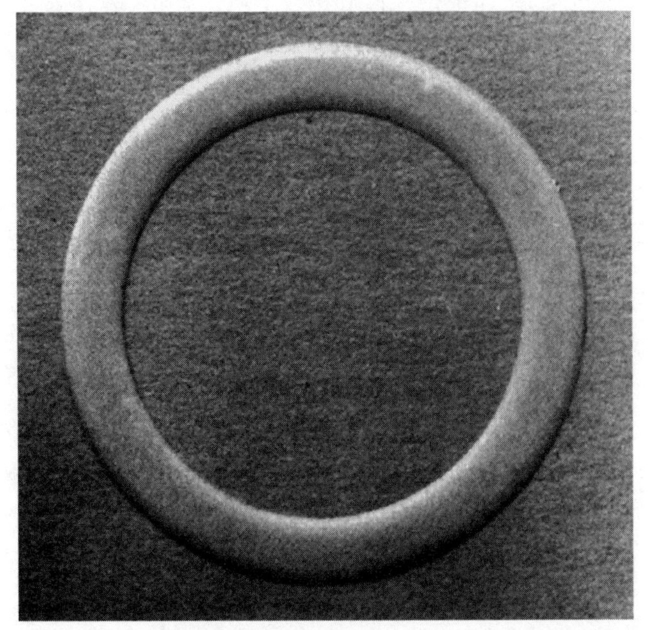

2-11 *Example of an inspection port. (Courtesy of Michael J. Kroes, William A. Watkins, and Frank Delp, Aircraft Maintenance and Repair, McGraw-Hill, 1993)*

an aluminum inspection plate which has an internal clip to hold it in position.

Dope

Aircraft dope's function is to shrink cotton or linen fabric and provide a watertight seal. The dope can be clear or pigmented. Silver is the most commonly used color because it blocks the sun's ultraviolet rays. Currently there are two types of dope in use.

One of the oldest finishes, *nitrate dope* is made with nitrocellulose mixed with plasticizers and thinners. The plasticizer usually gives the material flexibility by trapping the compound in the resin rather than creating a chemical bond. Ketone thinners hold the nitrocellulose in suspension and dry very quickly when exposed to air. In fact, the solvent flashes, or dries so rapidly by evaporation that it leaves behind a hard, flexible coating on the fabric. Because of the flammability, both while applying it and after it has dried, most users have switched to a less flammable dope.

Butyrate dope is made with cellulose acetate butyrate, mixed with plasticizers and a solvent. Butyrate dope is much more fire resistant than nitrate dope. One characteristic of Butyrate dope is the large amount of fabric shrinkage that occurs during application. For this reason, use care in the amount of dope applied to the fabric, because the shrinking action can actually bend or warp the supporting airframe structures.

Polyester fabrics are shrunk by heat rather than chemical action. For these fabrics a combination of nitrate and butyrate dopes have been produced. These provide good adhesion characteristics and do not shrink the fabric to any significant degree.

Fungicidal Agents

Organic fabrics, like cotton and linen, are subject to attack by fungus. Fungicidal additives are introduced into the doping process to help eliminate this problem. Two of the most commonly used fungicide are zinc dimethyl dithio carbonate and copper naphthonate. Make sure to follow the manufacturer's directions for use.

Covering the Aircraft

Most fabric manufactures have developed excellent guidelines to follow when doing a recovering project with their product. First, choose a fabric that has the same strength (or greater) as the original fabric. Grade A cotton and other organic fabrics have almost exclusively been replaced by the synthetic fibers. Two leading synthetic fabrics, Stits Poly-Fiber and Ceconite, have been approved by Supplemental Type Certificates (STC) for almost all aircraft manufactured with fabric skin. Fiberglass fabric such as Razorback is an FAA-approved fabric for use on any aircraft, regardless of the fabric previously employed. Razorback is so widely approved because it meets the requirements of FAA advisory circular 22-44. Other fiberglass products are available, but the builder or maintenance technician must determine if the products are airworthy. Checking the selvage on synthetic fabric is one way to determine if that material is approved and appropriate to your needs.

After carefully choosing the fabric, it's time to prepare the aircraft structure for the recovering process. Take the time required to carefully check over the aircraft frame for any damage and to determine the structural integrity of the machine. Make sure to protect any varnished or coated structures which may be damaged by any unintentional dripping or running of the dope or solvents. Complete all the interior work, such as installing control cables or pulleys, while the interior components are still easy to reach. Also, it's important to have laid out the location of the inspection ports and grommets before starting the actual recovering process.

Installing the Fabric

The blanket or envelope methods are used to cover an aircraft structure. The envelope method is far and away the easier of the two methods: A factory-manufactured envelope or sleeve slips right over the major structural areas of the aircraft, such as the wings and horizontal stabilizer. The blanket method, in contrast, uses a large piece of fabric, fashioning it to cover sections of the aircraft structure being recovered. Since this fabric comes off a large roll, it frequently has to be sewn together to get a piece large enough to fit. Once the pieces are sewn together and the fabric is on the structure, any seams will have to be closed by sewing them together. Occasionally a mechanical device or attachment can be utilized to hold the fabric in place on the aircraft structure. Once the fabric is secured in place, it can be shrunk to fit by using dope or heat, depending on the fabric and the process being employed. Rib lacing is a process which connects the fabric to the aircraft structure by a series of thread and knots. Again, it's important to follow the aircraft manufacturer's instructions. Once the final coats of dope have been applied, the aircraft is ready for the paint shop.

Care and Repair of Fabric-Covered Aircraft

Many pilots shy away from fabric airplanes, thinking that they will be more expensive and time-consuming to care for than a traditional aluminum airplane. However, with proper care fabric can give the owner many years of reliable and safe use. Unlike aluminum aircraft, testing the strength of the fabric is an important preventative measure for fabric aircraft and should be done during each annual inspection. Some of the most damaging elements to the fabric include sunlight, pollution, mildew, and spilled aircraft fluids. The organic fabrics are especially vulnerable to these conditions. When strength-testing the fabric, choose a location on the aircraft that is most susceptible to weakening or damage caused by environmental conditions. Remember that dark colors absorb more ultraviolet rays than light colors, so any darkly painted area would be a good testing site. Another good location to test the fabric is any low point in the fuselage where fuel or hydraulic fluid could collect. To test an area, use a Maule punch tester or a Seyboth tester (Fig. 2-12). The Seyboth actually punctures the fabric and the tensile strength is read off a scale on the instrument. The Maule punch tester doesn't actually

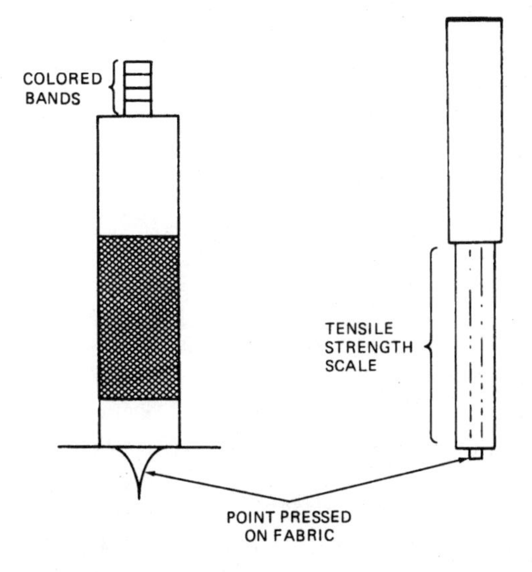

2-12 *A Seyboth tester (left) and a Maule-type tester (right). (Courtesy of Michael J. Kroes, William A. Watkins, and Frank Delp, Aircraft Maintenance and Repair, McGraw-Hill, 1993)*

penetrate the fabric but indicates how many pounds per inch of force is being applied directly to the fabric. Punch testers, while an excellent tool, are not the only reliable way to test fabric strength. In fact, the number of coats of dope actually affects the tester's reliability as a measuring tool. The most accurate way to test a fabric's strength is to remove a small portion and perform a tensile strength test. Here, the fabric is loaded onto a machine which grips each end of the fabric and pulls it apart until the load is equal to the minimum standard for that fabric or the fabric rips. The piece of fabric should be taken from an area of the aircraft considered the weakest, and all the dope should be removed. The minimum strength required depends on the type of fabric being tested. Grade A cotton, for example, should test to 56 lb/in. Polyester fabrics should test to the minimums of the fabric that they replaced. It's interesting to note that fiberglass fabrics don't need to be strength-tested, since they don't deteriorate under most normal environmental conditions. However, it is still important to check the fiberglass fabric for any type of damage or serious deterioration of the gel coat.

Repairs

Tears in fabric can normally be repaired depending on their location and severity. One should also consider the age and condition of the entire fabric covering to determine if a repair is warranted. Sometimes recovering the entire aircraft is the better option. A small rip is easily repaired with a baseball stitch (the stitchings resemble those found on a baseball). First, remove all the dope away from the rip. If used carefully, acetone is an excellent solvent. Once the area is clean, stitch the fabric back together using a baseball stitch. (See Fig. 2-13.) Make sure to lock-stitch the repair every 8 in. After sewing, use pinking shears to cut a piece of fabric large enough to cover the seam.

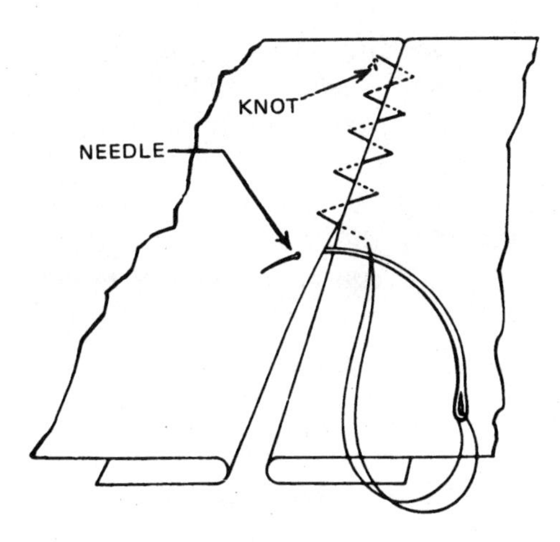

2-13 *Baseball stitch. (Courtesy of Michael J. Kroes, William A. Watkins, and Frank Delp, Aircraft Maintenance and Repair, McGraw-Hill, 1993)*

The patch should overlap the seam by at least 2 in in every direction from the tear. Dope the patch in place and sand lightly between coats until the patch blends with the original fabric.

Restoring Old Fabric

Sometimes the dope on an aircraft will harden with age and begin cracking. If it is determined that the fabric underneath the dope is sound, it is possible to rejuvenate it. Use a commercial rejuvenation product, which will soften and penetrate the old hard dope giving it new flexibility. Then seal any cracks or splits and finish up by applying a new coat of dope.

Structural Metals

Knowledge and understanding of the uses, strengths, limitations, and other characteristic structural metals is vital to properly construct and maintain any equipment, especially airframes. In aircraft maintenance and repair, even a slight deviation from design specification, or the substitution of inferior materials, may result in the loss of both lives and equipment. The use of unsuitable materials can readily erase the finest craftsmanship. The selection of the correct material for a specific repair job demands familiarity with the most common physical properties of various metals.

Properties of Metals and Explanation of Terms

Of primary concern in aircraft maintenance are such general properties of metals and their alloys as hardness, malleability, ductility, elasticity, toughness, density, brittleness, fusibility, conductivity contraction, and expansion, and so forth.

Hardness refers to the ability of a metal to resist abrasion, penetration, cutting action, or permanent distortion. Hardness may be increased by cold-working the metal and, in the case of steel and certain aluminum alloys, by heat treatment. Structural parts are often formed from metals in their soft state and then heat-treated to harden them so that the finished shape will be retained. Hardness and strength are closely associated properties of metals.

Brittleness is the property of a metal which allows little bending or deformation without shattering. A brittle metal is apt to break or crack without change of shape. Because structural metals are often subjected to shock loads, brittleness is not a very desirable property. Cast iron, cast aluminum, and very hard steel are examples of brittle metals.

A metal which can be hammered, rolled, or pressed into various shapes without cracking, breaking, or having some other detrimental effect, is said to be *malleable*. This property is necessary in sheet metal that is worked into curved shapes such as cowlings, fairings, or wingtips. Copper is an example of a malleable metal.

Ductility is the property of a metal which permits it to be permanently drawn, bent, or twisted into various shapes without breaking. This property is essential for metals used in making wire and tubing. Ductile metals are greatly preferred for aircraft use because of their ease of forming and resistance to failure under shock loads. For this reason, aluminum alloys are used for cowl rings, fuselage, and wing skin, and formed or extruded parts, such as ribs, spars, and bulkheads. Chrome molybdenum steel is also easily formed into desired shapes. Ductility is similar to malleability.

Elasticity is that property which enables a metal to return to its original shape when the force which causes the change of shape is removed. This property is extremely valuable because it would be highly undesirable to have a part permanently distorted after an applied load was removed. Each metal has a point known as the *elastic limit* beyond which it cannot be loaded without causing permanent distortion. In aircraft construction, members and parts are so designed that the maximum loads to which they are subjected will not stress them beyond their elastic limits. This desirable property is present in spring steel.

A material which possesses *toughness* will withstand tearing or shearing and may be stretched or otherwise deformed without breaking. Toughness is a desirable property in aircraft metals.

Density is the weight of a unit volume of a material. In aircraft work, the specified weight of a material per cubic inch is preferred, since this figure can be used in determining the weight of a part before actual manufacture. Density is an important consideration when choosing a material to be used in the design of a part in order to maintain the proper weight and balance of the aircraft.

Fusibility is the ability of a metal to become liquid by the application of heat. Metals are fused in welding. Steels fuse around 2600°F and aluminum alloys at approximately 1100°F.

Conductivity is the property which enables a metal to carry heat or electricity. The heat conductivity of a metal is especially important in welding because it governs the amount of heat that will be required for proper fusion. Conductivity of the metal, to a certain extent, determines the type of jig to be used to control expansion and contraction. In aircraft, electrical conductivity must also be considered in conjunction with bonding, to eliminate radio interference.

Contraction and expansion are reactions produced in metals as the result of heating or cooling. Heat applied to a metal will cause it to expand or become larger. Cooling and heating affect the design of welding jigs, castings, and tolerances necessary for hot-rolled material.

Selection Factors

Strength, weight, and reliability are three factors which determine the requirements to be met by any material used in airframe construction and repair. Airframes must be strong and yet as light in weight as possible. There are very definite limits to which increases in strength can be accompanied by increases in weight. An airframe so heavy that it could not support a few hundred pounds of additional weight would be of little use.

All metals, in addition to having a good strength: weight ratio, must be thoroughly reliable, thus minimizing the possibility of dangerous and unexpected failures. In addition to these general properties, the material selected for a definite application must possess specific qualities suitable for the purpose.

The material must possess the strength required by the dimensions, weight, and use. There are five basic stresses which metals may be required to withstand: tension, compression, shear, bending, and torsion.

The tensile strength of a material is its resistance to a force which tends to pull it apart. Tensile strength is measured in pounds per square inch (lb/in^2) and is calculated by dividing the load, in pounds, required to pull the material apart by its cross-sectional area, in square inches.

The compression strength of a material is its resistance to a crushing force which is the opposite of tensile strength. Compression strength is also measured in pounds per square inch.

When a piece of metal is cut, the material is subjected, as it comes in contact with the cutting edge, to a force known as *shear*. Shear is the tendency on the part of parallel members to slide in opposite directions. It is like placing a cord or thread between the blades of a pair of scissors. The shear strength is the shear force in pounds per square inch at which a material fails. It is the load divided by the shear area.

Bending can be described as the deflection or curving of a member due to forces acting upon it. The bending strength of material is the resistance it offers to deflecting forces. Torsion is a twisting force. Such action would occur in a member fixed at one end and twisted at the other. The torsional strength of material is its resistance to twisting.The relationship between the strength of a material and its weight per cubic inch, expressed as a ratio, is known as the strength:weight ratio. This ratio forms the basis for comparing the desirability of various materials for use in airframe construction and repair. Neither strength nor weight alone can be used as a means of true comparison. In some applications, such as the skin of monocoque structures, thickness is more important than strength, and, in this instance, the material with the lightest weight for a given thickness or gauge is best. Thickness or bulk is necessary to prevent buckling or damage caused by careless handling.

Corrosion is the eating away or pitting of the surface of the internal structure of metals. Because of the thin sections and the safety factors used in aircraft design and construction, it would be dangerous to select a material possessing poor corrosion-resistance characteristics.

Another significant factor to consider in maintenance and repair is the ability of a material to be formed, bent, or machined to required shapes. The hardening of metals by coldworking or forcing is termed *work-hardening*. If a piece of metal is formed (shaped or bent) while cold, it is said to be *coldworked*. Practically all the work an aviation mechanic does on metal is coldwork. Although this is convenient, it causes the metal to become harder and more brittle.

If the metal is coldworked too much, that is, if it is bent back and forth or hammered at the same place too often, it will crack or break. Usually, the more malleable and ductile a metal is, the more coldworking it can stand. Any process which involves controlled heating and cooling of metals to develop certain desirable characteristics (such as hardness, softness, ductility, tensile strength, or refined grain structure) is called *heat treatment* or *heat treating*. With steels the term *heat treating* has a broad meaning and includes such processes as annealing, normalizing, hardening, and tendering.

In the heat treatment of aluminum alloys, only two processes are included: (1) the hardening and toughening process, and (2) the softening process. The hardening and toughening process is called *heat treating,* and the softening process is called *annealing*.

Aircraft metals are subjected to both shock and fatigue (vibrational) stresses. Fatigue occurs in materials which are exposed to frequent reversals of loading or repeatedly applied loads, if the fatigue limit is reached or exceeded. Repeated vibration or bending will ultimately cause a minute crack to occur at the weakest point. As vibration or bending continues, the crack lengthens until the part completely fails. This is termed shock and fatigue failure. Resistance to this is known as shock and fatigue resistance. It is essential that materials used for critical parts be resistant to these stresses.

Owner Maintenance

Most pilots are concerned with the appearance of their aircraft, both inside and out; fortunately, FAR 43 allows the owner or pilot a lot of leeway in sprucing up the interior and the exterior. Many parts of the interior can be easily replaced and several manufacturers offer kits that turn an old interior new again.

Replacing Plastic Sidewall and Door Coverings

Replacing plastic sidewall and door coverings requires a screwdriver and pliers, and the following steps should serve as guidelines:

1. The first step is to remove the old plastic sidewall covering in the cockpit and cabin area. Most coverings are fastened to the aircraft structure by use of snaps; use the pliers to remove the snaps and remove the sidewall. (See Fig. 2-14.)
2. Clean out any debris or dirt that may have accumulated behind the panels. Be careful not to remove or damage any insulating material.

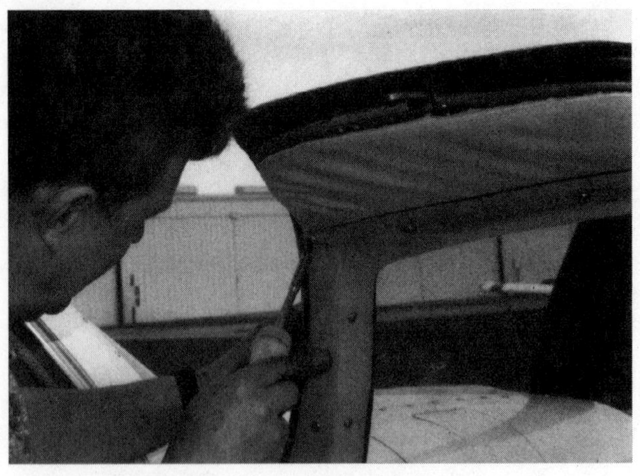

2-14 *Carefully remove plastic sidewall.*

3. Install the new plastic piece using new snap fasteners. These fasteners should fit into the holes previously used. Press the fasteners into position by hand.
4. Complete the logbook entry.

Replacing Cockpit and Cabin Carpeting

Follow these steps to replace carpeting:

1. Most aircraft carpeting is held into position by the use of snaps. Simply unsnap the carpet and remove.
2. Inspect the snap receptacles for any damage. If damaged, use a screwdriver to remove and replace with a new one.
3. Lay out the new carpet and align the snaps into position. Press firmly and lock the carpet into position (Fig. 2.15). Be sure that the new carpet does not interfere with the seat tracks.
4. Enter into the logbook. (See Chap. 8.)

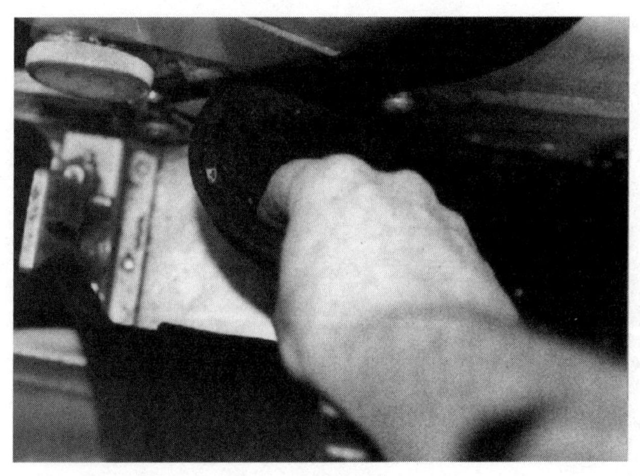

2-15 *Snap carpet into place.*

Replacing Seatbelts and Windows

Seatbelts become frayed and worn over time. Replacement belts are readily available and easy to install. Make sure that the replacement belt is designed and approved for the installation.

Using an open-ended wrench, follow these steps to replace seatbelts:

1. Remove the bolt that holds the lap belt to the seat frame or floor of the aircraft. Inspect the anchoring bolt for any damage. If damaged, replace with a new bolt of the same type (Fig. 2-16).
2. Thread the seatbelt webbing through the securing hardware.
3. Reinstall the anchoring bolt to the floor or seat frame and tighten.
4. Repeat this process on the shoulder harness belt, if required.
5. Enter the replacement into the aircraft logbook. (See Chap. 8.)

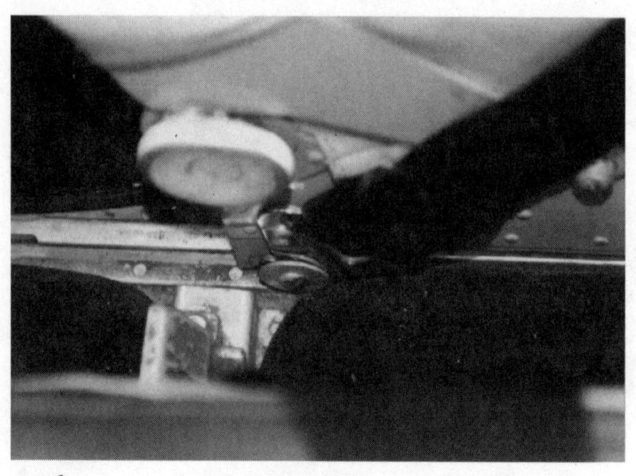

2-16 *Remove bolt holding seatbelt to frame.*

As windows age, they tend to get hazy or cracked or crazed. Owners or pilots are allowed to replace side windows as long as the replacement operation does not interfere with any controls. The job can be simple; however, pay close attention to how the window comes out so reinstalling it is done correctly. Tools needed are a screwdriver and an open-ended wrench. Follow these steps:

1. Remove any plastic or other material used as cover or molding from around the window frame. This will expose the brackets holding the window in place.
2. Using an open-ended wrench, loosen and remove the small bolts holding the brackets to the frame. The window should now slide out of the frame. Be alert to any gaskets or sealing material around the window.
3. Clean the frame area where the replacement window fits. Pay particular attention to any broken brackets. Replace as necessary.
4. Peel the protective covering from just around the edges of the new window. Leave the

majority of the covering intact to protect the window during the installation. If the window requires a new gasket or sealing material, install it now.

5. Reinstall the window in the frame and center. Once correctly positioned, tighten the brackets which hold the window to the frame.
6. Replace the decorative moldings and remove the remainder of the protective film from the window. Clean the window with a soapy solution.
7. Enter into the aircraft logbook. (See Chap. 8.)

3

Understanding and Repairing Electrical Systems

Definitions

We know that electricity moves at the speed of light or 186,000 miles per second. Although true, this statement is not entirely accurate. Instead, consider the force of electricity moving at the speed of light. As an example, turn on a fuel pump in an airplane. A single electron does not flow from the fuel pump switch, accelerate to the speed of light, and energize the fuel pump. (Surprisingly, individual electrons move rather slowly between atoms, but the cumulative effect is lightning fast.) Take the fuel pump switch as an example. The conductor, in this case a piece of wire, is running between the switch and the pump. The wire is packed to capacity with electrons. Imagine the wire as a clear plastic tube full of marbles. (See Fig. 3-1.) In fact, it's so packed, another marble could not be stuffed in the tube. That's the resting condition of the wire between the switch and

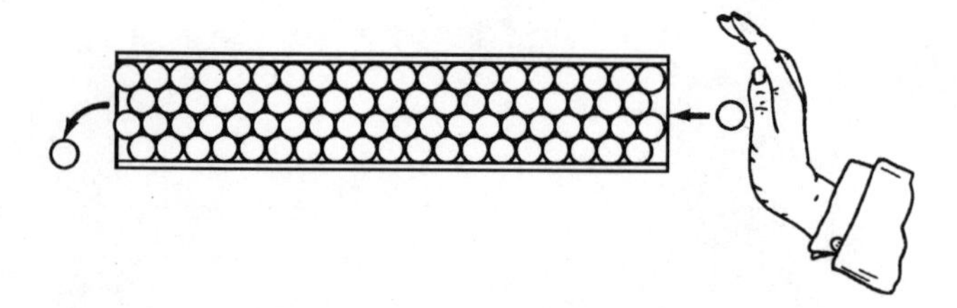

3-1 *Demonstration of current flow. (Courtesy of Thomas K. Eismin, Aircraft Electricity and Electronics, 5th ed., McGraw-Hill, 1995)*

the fuel pump. Of course, the wire is full of electrons, not marbles. When the fuel pump switch is turned on, the effect is similar to forcing another marble into the tube. What happens? A marble immediately falls out of the other end. This is current. Current is the flow of electrons.

It helps to think of current using a water analogy. Just like water, current is measured by quantity and rate of flow. In electrical circuits, a *coulomb* is a unit of quantity. One coulomb is equal to the amount of electricity that causes 0.001118 of a gram of silver to be deposited upon one electrode as it passes through a silver nitrate solution. An electrode is just a terminal in an electrical circuit. A more useful definition of a coulomb is 1 coulomb equals 6.28 billion billion electrons. In order to measure the rate of an electrical current, the number of coulombs per second passing a defined point in a circuit is tabulated. Of course, this number would be extremely cumbersome, so instead of calling the rate of flow coulombs per second, the term *ampere* is used. One ampere is the rate of flow of 1 coulomb per second. Current is measured in amperes, or sometimes just *amps* for short. So, what induces electricity to flow? What creates current? Again, go back to the water scenario. Electricity, like water, flows downhill. To be more precise, it flows from a high pressure to a low pressure. Think of a 55-gallon drum located high on a hill and full of water, connected to a 5-gallon bucket (located far down the hill) by a piece of pipe. (See Fig. 3-2.) Imagine that the water in the 55-gallon drum was electrons, it's easy to see how the flow from the large number of electrons to the smaller number of electrons occurs. Now imagine a valve in the pipe connecting the two water containers. With the valve closed, there is a difference in potential between the two tanks. If the valve is opened, the water or electrons would flow from the point of high potential to the point of low potential. Just think of the water as electrons. This electron-moving potential is called an *electromotive force*. A *volt* is the term used to measure this electromotive force, or, put another way, a volt measures the difference between high potential and low potential. One volt is the electromotive force required to make current flow at the rate of one ampere.

Voltage is just electrical pressure. An airplane battery stores this pressure, and when given a release it creates a force that moves electrons from one place to another. An airplane alternator or generator also creates this pressure, or force, to move electrons. Of course, there is resistance to the flow of an electrical current. This resistance is also measurable and is called an *ohm*. Think of a pipe. If the inside of the

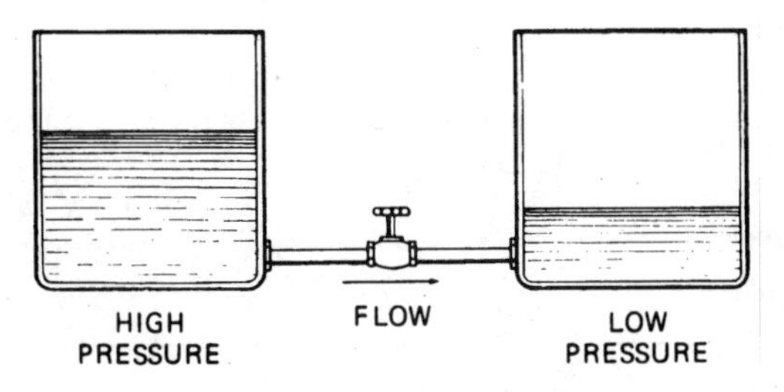

3-2 *Difference of pressure. (Courtesy of Thomas K. Eismin, Aircraft Electricity and Electronics, 5th ed., McGraw-Hill, 1995)*

pipe were caked with lime deposits, the flow of water at a given pressure would be less than if the pipe were clean and unrestricted. So it is with electrical currents. Of course, electrical resistance concerns material that will not accept extra electrons very easily, material considered highly resistant. What makes a substance a good conductor or bad is related to the number of electrons in the valence orbit of the material's atoms. (See Fig. 3-3.) A valence orbit is the outermost orbit of an atom. The electrons circling the atom in the valence orbit are called *valence electrons*. All atoms want their valence orbit full of electrons. If the orbit isn't full, the atom will easily accept more electrons. As a general rule, atoms with fewer than four valence electrons are called *conductors*. Gold and silver only have one valence electron, so they make excellent, although expensive, conductors. Semiconductors have exactly four valence electrons. In their natural state, a semiconductor, such as silicon, has a very high resistance to current flow, but change the number of electrons, and the resistance drops to a very low level. Atoms with more than four valence electrons are said to be *insulators*, that is, they resist current flow. Some common insulators are air, plastic, rubber, and fiberglass. Remember, just

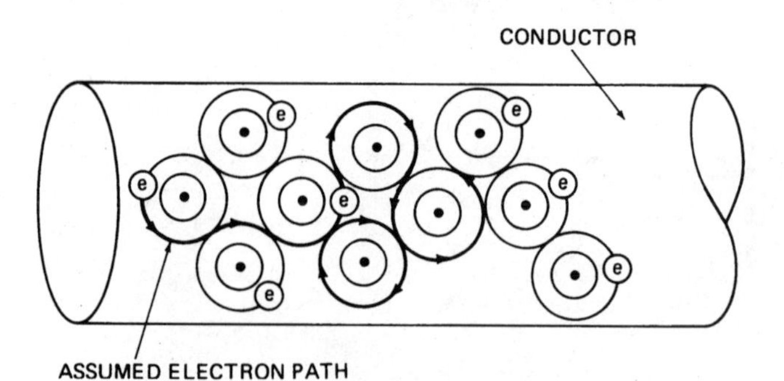

3-3 *Movement of electrons. (Courtesy of Thomas K. Eismin, Aircraft Electricity and Electronics, 5th ed., McGraw-Hill, 1995)*

3-4 *Typical aircraft alternator. (Courtesy of Thomas K. Eismin, Aircraft Electricity and Electronics, 5th ed., McGraw-Hill, 1995)*

because a material is a good conductor doesn't mean current will begin to flow. Something has to move those electrons, and in an aircraft, that something is the battery or alternator. (See Fig. 3-4.)

Batteries

The most common type of battery found in general aviation aircraft is the lead acid battery. (See Fig. 3-5.) Of lead acid batteries, two types are currently in use. The vented cell, which is the most common, is usually identified by the removable plugs on top. Simply unscrew the plugs to look directly inside the battery. A newer type of battery is the sealed lead acid battery, which has no removable top and is similar to the type of battery found in the family car. Both these batteries are secondary cell batteries, meaning that the chemical process that produces the electrical current can be reversed. Simply stated, the battery can be recharged. The battery in a light aircraft is primarily used for starting the engine. It also functions as an emergency backup should the electrical generator or

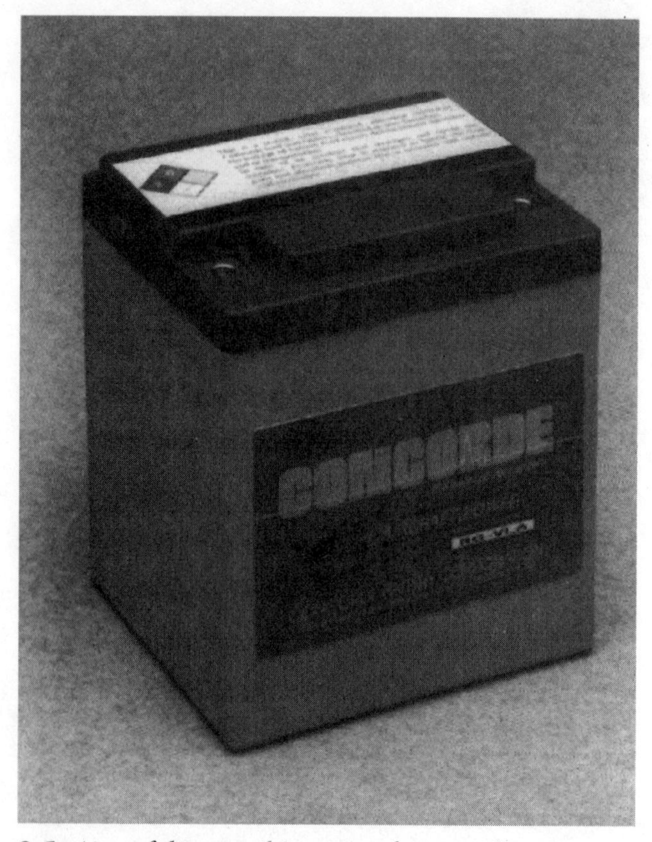

3-5 *Aircraft battery. (Courtesy of Concord Battery Company)*

alternator fail. A fully charged battery will normally last approximately 30 minutes depending on the power draw. For this reason, it's important to maintain the battery just like any other piece of emergency equipment.

Battery Maintenance Procedures

To ensure that the battery will be available in an emergency, use the following maintenance schedule.

Remember, when working with an aircraft battery, always wear safety glasses.

1. Always remove the negative lead first and replace it last. This will prevent arcing if the positive lead is inadvertently touched to the positive lead.
2. Be careful where tools are placed. A wrench accidentally touching both poles can short-circuit the battery and give a nasty burn in the process.
3. Do not jump-start an airplane from a car or any other external source. A dead battery can take several hours to recharge. In flight, a dead battery results in the loss of the electrical system backup. In addition, jump-starting allows a strong current surge into the airplane's battery and could damage its cells.
4. Inspect the battery cables to ensure tight and corrosion-free connections. Don't allow the cables to rub on anything which could wear a hole into the insulation.
5. Corrosion on the terminals should be removed with a baking soda solution and a stiff brush. Do not use a wire brush because of the danger of accidentally shorting out the battery by touching both terminals simultaneously.
6. Clean any residual electrolyte off the top of the battery with a baking soda solution. Spilled electrolyte is a great conductor and may discharge the battery. Ensure none of the soda solution gets into the battery's cells. It will neutralize the electrolyte, producing a dead battery.
7. Always service the battery with distilled water, not additional electrolyte.
8. Clean the battery with baking soda and water. Be sure to thoroughly flush the mixture with clean water and dry the battery before

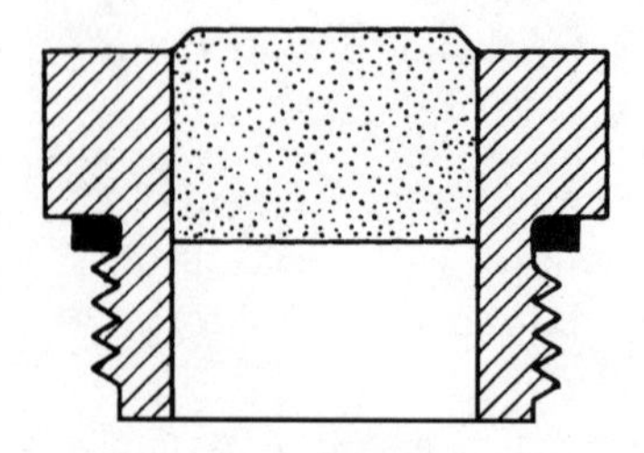

3-6 *Battery cap. (Courtesy of Thomas K. Eismin, Aircraft Electricity and Electronics, 5th ed., McGraw-Hill, 1995)*

replacing it in the aircraft. Clean the battery caps by soaking them in hot water—this cleans out their vent holes. (See Fig. 3-6.)

9. Inspect the battery box for any leakage or corrosion. Check for proper ventilation. If the box has corroded through it will have to be replaced. Inspect for any airframe damage below or near the battery box. Check on the fuselage externally where the battery overflow tube is located. Check for any leakage or corrosion.

See Table 3-1 for a troubleshooting guide for battery problems.

Battery Charging

Batteries are recharged by introducing a current in the opposite direction of discharge. (See Fig. 3-7.) While the engine is in operation, the alternator or generator performs this function and provides the battery with a constant voltage charge. A constant voltage charger works on the following principles: A 12-volt battery requires a 14-volt alternator or a 14-volt electrical charger to charge the battery. Remember a higher voltage flows to lower voltage. Hence, a 24-volt system needs a 28-volt charger. When a battery becomes fully charged, its amp reading will be slightly lower than the

Table 3-1. Battery Troubleshooting Guide

Observation	Probable cause	Corrective action
High trickle charge—when charging at constant voltage of 28.5 (±0.1) V, current does not drop below 1 amp after a 30-minute charge	Defective cells	While still charging, check individual cells. Those below 0.5 V are defective and should be replaced. Those between 0.5 and 1.5 V may be defective or may be unbalanced, those above 1.5 V are all right.
High trickle charge after replacing defective cells, or battery fails to meet amp-hour capacity check	Cell imbalance	Discharge battery and short-out individual cells for 8 hours. Charge battery using constant current method. Check capacity and if O.K., recharge using constant current method.
Battery fails to deliver rated capacity	Cell imbalance or faulty cells	Repeat capacity check, discharge and constant-current charge a maximum of three times. If capacity does not develop, replace faulty cells.

(*Cont.*)

Table 3-1. (*Continued*)

Observation	Probable cause	Corrective action
No potential available	Complete battery failure	Check terminals and all electrical connections. Check for dry cell. Check for high trickle charge.
Excessive white crystal deposits on cells (there will always be some potassium carbonate present due to normal gassing)	Excessive spewage	Battery subject to high-charge current, high temperature, or high liquid level. Clean battery, constant-current charge, and check liquid level. Check charger operation.
Distortion of cell case	Overcharge or high heat	Replace cell.
Foreign material in cells—black or gray particles	Impure water, high heat, high concentration of KOH, or improper water level	Adjust specific gravity and electrolyte level. Check battery for cell imbalance or replace defective cell.

Excessive corrosion of hardware	Defective or damaged plating	Replace parts.
Heat or blue marks on hardware	Loose connections causing overheating of intercell connector or hardware	Clean hardware and properly torque connectors.
Excessive water consumption; dry cell	Cell imbalance	Proceed as above for cell imbalance.

SOURCE: Thomas K. Eismin, *Aircraft Electricity and Repair*, 5th ed., New York: McGraw-Hill, 1995.

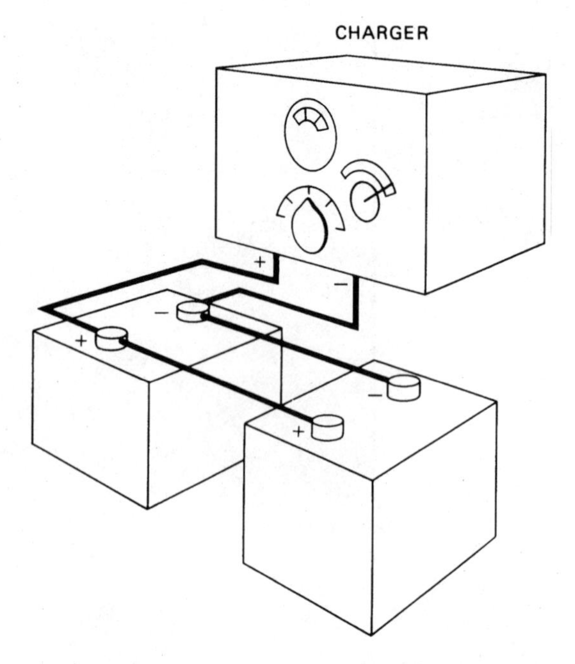

3-7 *Charging multiple batteries. (Courtesy of Thomas K. Eismin, Aircraft Electricity and Electronics, 5th ed., McGraw-Hill, 1995)*

charging unit. In addition, the charging current will have fallen off to approximately 1 amp. With a small flow of current, the battery can remain on the charger indefinitely, providing the battery has sufficient electrolyte. Of course, whenever recharging a battery, always observe these simple precautions:

1. Always recharge the battery in a well-ventilated, preferably open area.
2. Turn off the charger prior to disconnecting the battery.
3. When removing the battery from the aircraft, disconnect the negative lead first. When replacing the battery, hook up the negative side last.
4. Remove the caps from the battery during its recharging.

5. Always wear safety glasses and avoid getting electrolyte on exposed skin. Remember it's acid!

Battery Freezing

Lead-acid batteries exposed to cold temperatures are subject to plate damage due to freezing of the electrolyte. The freezing point of electrolyte for various specific gravity levels is shown in Table 3-2. To prevent freeze damage, maintain the specific gravity at a reasonably high level, bearing in mind that lead-acid batteries are subject to a constant discharge due to the internal chemical action. Nickel-cadmium battery electrolyte is not as susceptible to freezing because no appreciable chemical change takes place between the charged and discharged state. However, the electrolyte will freeze at approximately −75°F.

Table 3-2. Freezing Points for Different States of Charge in a Battery

Specific gravity	°F(°C)
1.300	−95 (−70.6)
1.285	−85 (−65)
1.275	−80 (−62.2)
1.250	−62 (−52.2)
1.225	−35 (−37.2)
1.200	−16 (−26.7)
1.175	−4 (−20.0)
1.150	+5 (−15.0)
1.125	+13 (−10.6)
1.100	+19 (−7.2)

Lead-acid batteries manufactured in the United States are considered fully charged when the specific gravity reading is between 1.275 and 1.300. A one-third-discharged battery reads about 1.240, and a two-thirds-discharged battery will show a specific gravity reading of about 1.200 when tested by a hydrometer. However, to determine precise specific gravity readings, a temperature correction should be applied to the hydrometer indication. As an example, the hydrometer reading is 1.260; the temperature of the electrolyte is 40°F, or 16 degrees below the norm established for battery electrolyte. Therefore, the corrected specific gravity reading of the electrolyte is 1.244. Take care to ensure that the electrolyte is returned to the cell from which it was extracted. When a specific gravity difference of 0.050 or more exists between cells of a battery, the battery is approaching the end of its useful life, and replacement should be considered. Electrolyte level may be adjusted by the addition of distilled water.

Replacing a Battery

Here's how to install a new battery in order to achieve long dependable service. Most new batteries are dry-shipped. That is, the electrolyte comes in a separate container. Wear eye protection and carefully open the sealed electrolyte container. Fill each battery cell until approximately ⅜ in of electrolyte covers the plates. Wait approximately an hour, then refill the cells to the recommended level. The battery can then be placed into service. However, better results can be achieved by leaving the battery on a charger for 15 hours. Be sure to clean up any spilled electrolyte.

Caution: Serious burns will result if the electrolyte comes in contact with any part of the body. Use rubber gloves, rubber apron, and protective goggles when handling electrolyte. If sulphuric acid is splashed on the body, neutralize with a solution of baking soda and water and shower or flush the affected area with water. For the eyes, use an eye fountain and flush with an abundance of water.

If potassium hydroxide contacts the skin, neutralize with 3% acetic acid, vinegar, or lemon juice and wash with water.

For the eyes, wash with a weak solution of boric acid or a weak solution of vinegar and flush with water.

Electrical Wires

There are literally hundreds or even thousands of feet of wire in the average light aircraft. Aircraft wire is exposed to severe environmental conditions. Because of this environment, all aircraft wires should be inspected frequently for cuts, abrasions, and corrosion. While building an aircraft or just replacing existing wires, it's imperative to choose the correct type of wire based on the current requirements, operating temperatures, and the wires' location in the aircraft. To determine which type of wire is most suitable, consider that the size of the wire must be sufficient to prevent any excessive voltage drop. It must also be able to withstand the current flow without overheating. In order to meet these two requirements, the following information is required:

1. The length of wire needed in feet
2. The number of amps the wire will be carrying
3. The amount of voltage drop allowed
4. If the current the wire is to carry is continuous or intermittent

Once these items are known, simply refer to Fig. 3-8. Of course, some precautions should be taken when installing or replacing wire. First, avoid splicing wires together. Never splice within 12 in of a terminal

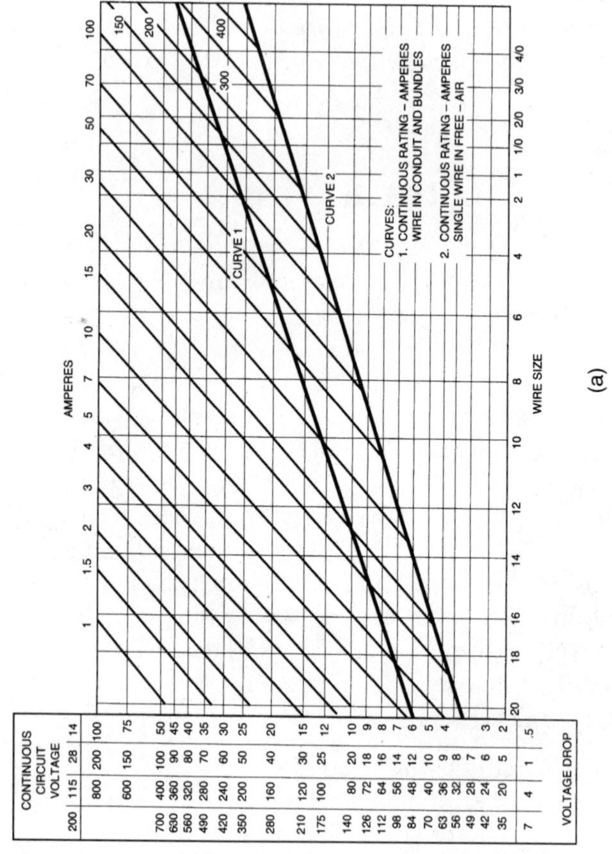
CONTINUOUS CIRCUIT VOLTAGE
200
115
28
14
WIRE LENGTH IN FEET/VOLT DROP
VOLTAGE DROP
AMPERES
WIRE SIZE
CURVE 1
CURVE 2
CURVES:
1. CONTINUOUS RATING – AMPERES WIRE IN CONDUIT AND BUNDLES
2. CONTINUOUS RATING – AMPERES SINGLE WIRE IN FREE – AIR

(a)

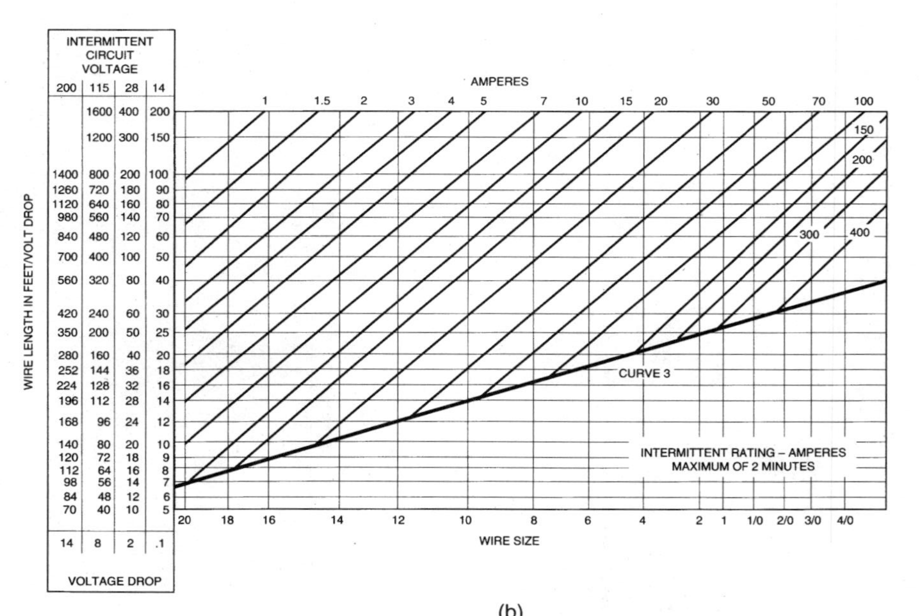

(b)

3-8 *Wire selection chart. (a) Continuous loads. (b) Intermittent loads. (Courtesy of Thomas K. Eismin, Aircraft Electricity and Electronics, 5th ed., McGraw-Hill, 1995)*

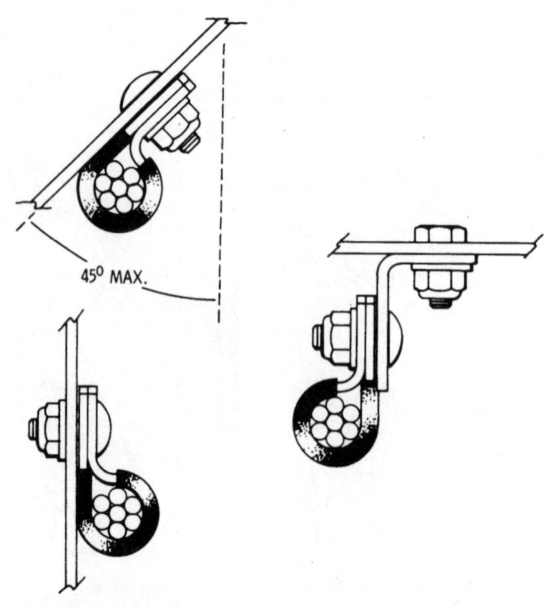

3-9 *Correct method for installing clamps. (Courtesy of Thomas K. Eismin, Aircraft Electricity and Electronics, 5th ed., McGraw-Hill, 1995)*

or termination device. Do not splice just to save a piece of scrap wire. Avoid splicing in areas subject to vibrations; vibrations frequently will cause a splice to come apart. When splicing wires, use an insulated splice connector. If unable to get an insulated connecter, cover the splice with a plastic sleeve secured at both ends of the splice. Never use electrical tape or adhesive-type tapes. Adhesive tapes become brittle with age and will fall off. Once the wire is run, secure it to a support structure and protect it from chafing and heat. When clamping a wire or wire bundle, use clamps that have nonmetallic inserts to protect the wires. You can tie the wire together between clamps, but do not tie the wires to any aircraft structure. Wires that are allowed to come in contact with the aircraft structure or other parts may chafe a hole in the insulation, leading to short circuits and inoperative equipment. Wires should always be supported or clamped if they are in close proximity to an aircraft structure which may damage them. (See Fig. 3-9.) Do not clamp the wires on or near any flammable fluid lines such as fuel, oil, hydraulic, or alcohol lines. Any type of break in the lines could lead to a serious and deadly fire. If running wires near hot equipment such as the exhaust stack or cabin heater, use the

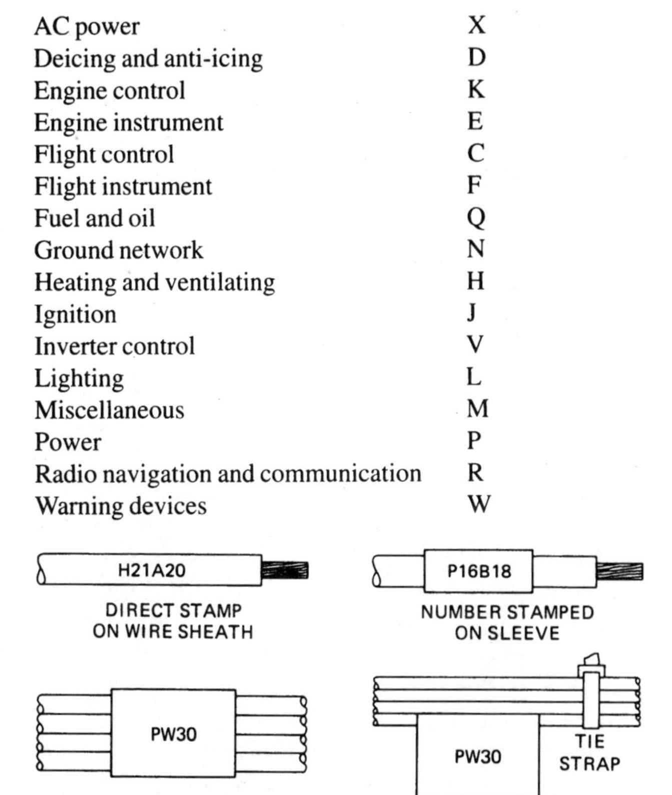

AC power	X
Deicing and anti-icing	D
Engine control	K
Engine instrument	E
Flight control	C
Flight instrument	F
Fuel and oil	Q
Ground network	N
Heating and ventilating	H
Ignition	J
Inverter control	V
Lighting	L
Miscellaneous	M
Power	P
Radio navigation and communication	R
Warning devices	W

3-10 *Methods of marking wires. (Courtesy of Thomas K. Eismin, Aircraft Electricity and Electronics, 5th ed., McGraw-Hill, 1995)*

correct insulation material. Asbestos or teflon are excellent choices for high heat areas.

Identification of Wires

All wires are marked with a wire identification number to help facilitate identification and maintenance. (See Fig. 3-10.) These wire identification marks are normally spaced 12 to 15 in apart and at the end and beginning of the wire. If the wire is short, the identification band is in the middle of the wire. A combination of letters and numbers identify each wire, what circuit it is part of, and the gauge size. The following chart can be used to identify wires by their function:

E — engine instruments wiring
J — ignition wiring
P — electrical power wiring
R — radios
M — miscellaneous wiring

Switches

A switch makes or breaks an electrical circuit. It turns an appliance on or off. One of the most used switches in an aircraft is the master switch. (See Fig. 3-11.) Almost everything electrical in the aircraft is routed through this switch. To obtain aircraft electrical power, it must be turned on. Activating the master allows the aircraft battery to energize the electrical buses. A *bus* is merely a power distribution point. Almost everything electrical in an aircraft is routed through a bus. One of the few electrical items that is not on a bus is the starter relay. It requires a tremendous amount of current to start an aircraft engine, and rather than making a large and heavy bus capable of withstanding that type current, most manufacturers bypass the electrical bus and energize the starter relay directly from the battery. In this type configuration, the master switch is used to energize the starter relay. Aircraft switches should always be installed so

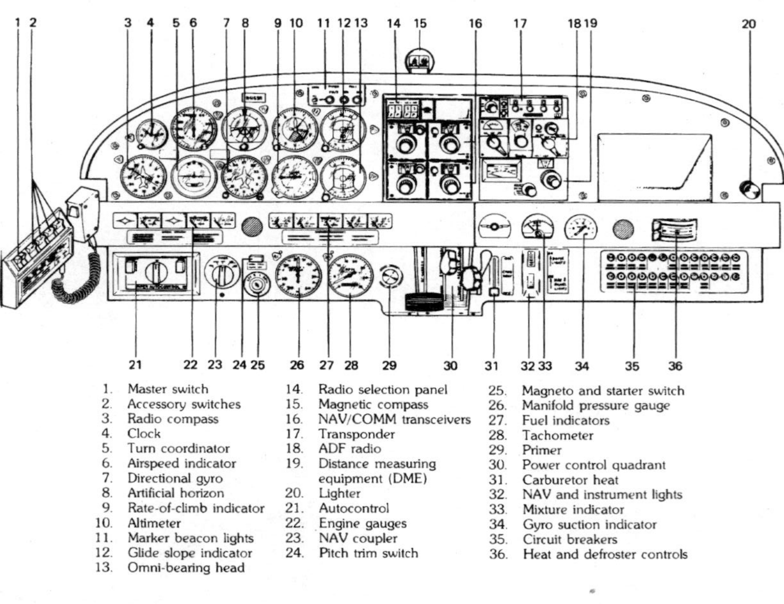

3-11 *Typical instrument panel with master switch. (Courtesy of Thomas K. Eismin, Aircraft Electricity and Electronics, 5th ed., McGraw-Hill, 1995)*

that the pilot will move the switch in the correct direction for the particular results desired.

Magnetism

No explanation of aircraft electrical systems is possible without a thorough understanding of magnetism. A magnet attracts iron or steel. Hold a piece of metal to a magnet and an immediate attraction occurs. The metal is attracted because the magnet has a magnetic field surrounding it. This magnetic field is composed of invisible lines of force that flow from the north pole of the magnet and travel to the south pole. (See Fig. 3-12.) These lines of force will never intersect each other, simply because like poles repel each other and opposite poles attract.(See Fig. 3-13.) This almost palatable force is called *magnetic flux* as it travels from the north pole to the south pole of a magnet. The flow of this force from one pole to another is called a *magnetic circuit*. Placing an iron bar across the two poles allows the magnetic circuit to flow through the bar, because it offers less resistance to circuit flow than the air.

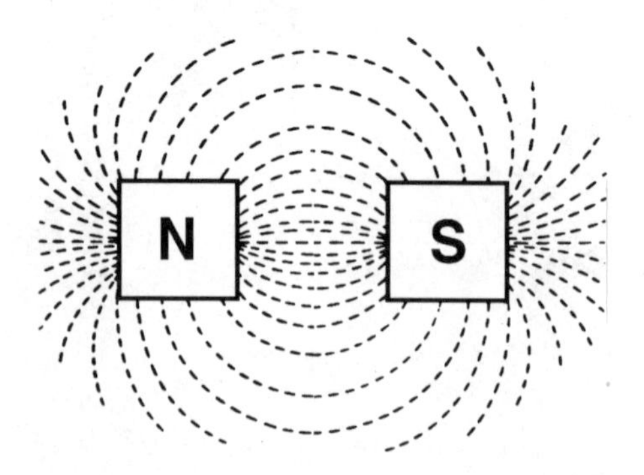

3-12 *A magnetic field. (Courtesy of Thomas K. Eismin, Aircraft Electricity and Electronics, 5th ed., McGraw-Hill, 1995)*

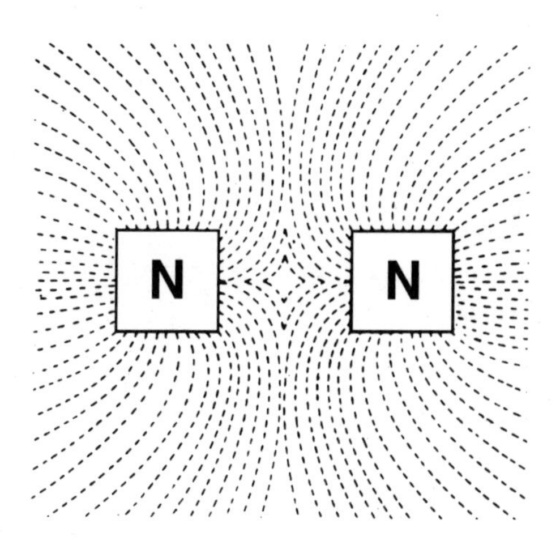

3-13 *Like poles repel each other. (Courtesy of Thomas K. Eismin, Aircraft Electricity and Electronics, 5th ed., McGraw-Hill, 1995)*

Electromagnets are created by using an electrical current to create a magnetic field. All conductors and wires that carry current have a magnetic field around them. This field is created by the movement of electrons passing through the conductor. Usually, this magnetic field is very weak, but by forming the conductor into a coil, the magnetic field increases in strength. (See Fig. 3-14.) A coil is nothing more than a piece of iron core wrapped with wires that are carrying current. Note that anytime a looped wire is carrying current, it will assume some of the properties of a magnet. That is, the wire will have a north and a south pole. If these wires are wrapped around a soft piece of iron, a magnet will be created because the magnetic field finds it easier to travel through the iron than travel through the air. A typical electromagnet is created by taking an insulated wire and wrapping it numerous times around a soft iron core. The strength of the electromagnet is directly proportional to the current flowing through the wrapped

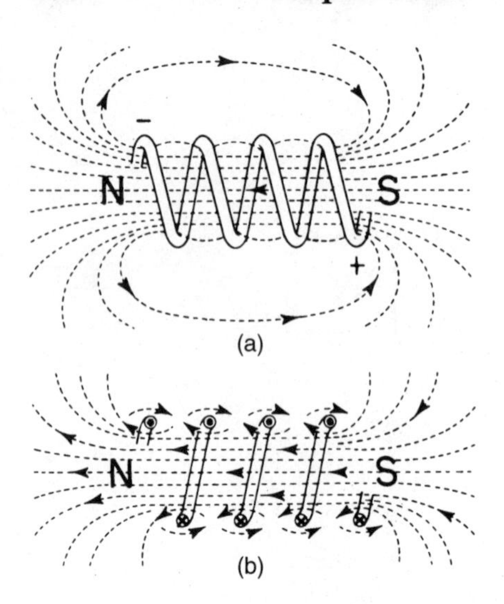

3-14 *(a) Magnetic field of a coil, or multilooped wire. (b) The lines of force form a pattern through all the loops, causing a high concentration of flux lines. (Courtesy of Thomas K. Eisman, Aircraft Electricity and Electronics, 5th ed., McGraw-Hill, 1995)*

wire and to the number of times the wire has been wrapped around an iron bar. By increasing the current or wrapping more wires around the coil, an increase in the strength of the electromagnet results. That is the basic principle of an electromagnet.

The next step is generating usable electrical power from the electromagnet. Whenever electrical energy is transferred by using a magnetic field, the process is called *electromagnetic induction*. (See Fig. 3-15.) Relative movement between a wire and a magnetic field creates electromagnetic induction. This relative movement can be created one of two ways: The first is by having a stationary conductor and a moving magnetic field. The other is a stationary field and a moving conductor. Typically, this is accomplished by rotating a coil inside a magnetic field or rotating a magnetic field inside a wire coil. These are the basic principles that explain how the alternator, generator, and magnetos function.

Magnetos

A magneto ignition system was developed to provide a dependable ignition system that is completely independent of the aircraft's electrical system. Although there are several types of magneto systems, most general aviation aircraft use a rotating magnet system.

Alternators

The most common type of alternator found on light aircraft is the AC alternator. Most units are manufactured

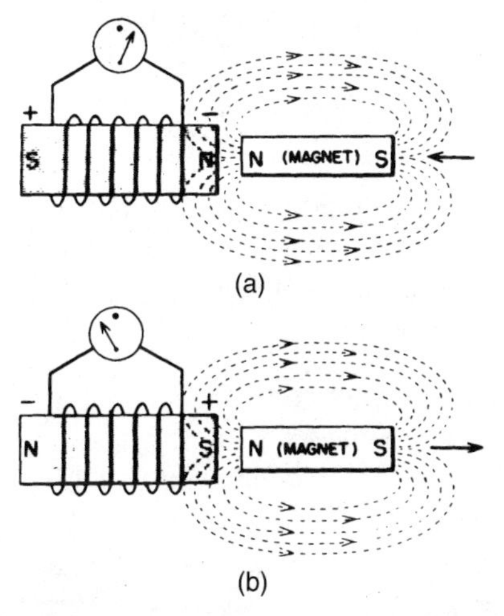

3-15 *(a) Direct current flowing through a coil, causing the core to become magnetized with the same polarity. (b) Current is reversed, thereby reversing the polarity. (Courtesy of Thomas K. Eismin, Aircraft Electricity and Electronics, 5th ed., McGraw-Hill, 1995)*

by Ford Motor Company, Chrysler, or Delco-Remy. To understand how an alternator works, it is necessary to know something about the principles of AC generation. Briefly, any time a conductor is cut by magnetic lines of force, a voltage will be induced into the conductor, and the direction that voltage takes will depend on the direction of magnetic flux and the direction of movement across the flux.

Let's apply those principles to an alternator and see if we can't make sense of it: First, you're going to need to know a few terms.

Stater. These are the stationary windings of an alternator. Located inside the alternator, these three separate windings are located 120 degrees apart.

Rotor. This is the rotating part of the alternator. It rotates inside the stater. A rotor usually will have 8 or 12 poles that alternate between north and south polarity. As the rotor spins inside the stater, the rotor creates a rotating field.

An alternator is a machine that converts a mechanical energy into electrical energy by means of electromagnetic induction. Alternators are classified by one of the following methods.

1. A direct connected, direct current (dc) system. This type of alternator uses direct current from the battery to excite the alternator field, after which the alternator is self-excited.
2. A transformation and rectification alternating current (ac) system. This method uses residual magnetism for initial ac voltage buildup, after which the field is supplied with rectified voltage from the alternator.
3. Integrated brushless type system. This system has a dc generator on the same shaft as an alternator. The excitation circuit is completed through silicon rectifiers rather than a

commmutator and brushes. The rectifiers are mounted on the generator shaft, and their output is fed directly to the alternator's rotating field.

Another method of classifying alternators is by the number of phases of voltage output. A three-phase or polyphase circuit is used in most general aviation aircraft alternators instead of a single or two-phase alternator. The three-phase alternator has three single-phase windings spaced so that the voltage induced in each winding is 120 degrees out of phase with the voltage in the other two windings.

Circuit Protection

Fuses

Most circuits fail because of a short somewhere in the wiring. (See Fig. 3-16.) A short circuit occurs when the insulation is rubbed off a wire or a clamp isn't properly insulated and the wire rubs against the aircraft structure resulting in a break in the insulation on the wire. Current then flows from the break in the wire back to the power source. The power source can be either the battery or alternator. As excessive current flows through the break, the wire becomes hot and may burn off more insulation or result in a

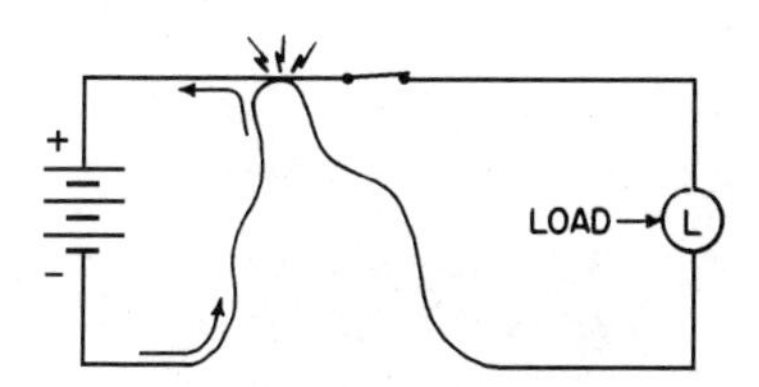

3-16 *A diagram of a short circuit. (Courtesy of Thomas K. Eismin, Aircraft Electricity and Electronics, 5th ed., McGraw-Hill, 1995)*

fire. Many fires are thwarted, however, by a fuse or a circuit breaker. A fuse is nothing more than a slim strip of wire designed to melt at a preset temperature. It's normally made of lead or a lead and tin mixture or some other type of low-melting-point metal. When the current flowing though a fuse exceeds its capability, the fuse melts and breaks the circuit. Usually fuses are enclosed in a glass tube or some other type of heat-resistant material. An adequate supply of replacement fuses should always be on board the aircraft.

Circuit Breaker

A circuit breaker performs the same function as a fuse but is resettable. (See Fig. 3-17.) Most circuit breakers trip when excessive temperature acts on a bimetallic strip. These breakers cannot be reset until the temperature returns to a normal range. As a pilot, wait several minutes before attempting to reset a tripped circuit breaker once the overload has been rectified.

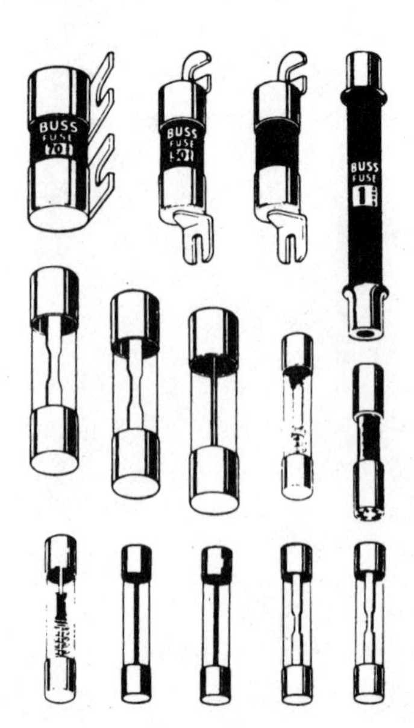

3-17 *Typical fuses. (Courtesy of Thomas K. Eismin, Aircraft Electricity and Electronics, 5th ed., McGraw-Hill, 1995)*

Electrical Systems

Care of Electrical Systems

The satisfactory performance of an aircraft depends on the continued reliability of the electrical system. Reliability of the system is usually proportional to the amount of maintenance received and the knowledge of those who perform such maintenance. It is, therefore, important that maintenance be accomplished using the best techniques and practices to minimize the possibility of failure. This handbook is not intended to supersede or replace any government specification or specific manufacturer's instruction regarding electrical system maintenance.

The term *electrical system* as used in this book means those parts of the aircraft which generate, distribute, and utilize electrical energy, including their support and attachments.

Inspection and Operation Checks

Frequently inspect equipment, electrical assemblies, and wiring installations for damage, general condition, and proper functioning to ensure the continued satisfactory operation of the electrical system. Adjust, repair, overhaul, and test electrical equipment and systems in accordance with the recommendations and procedures set forth in aircraft and/or component maintenance instructions. Replace components of the electrical system that are damaged or defective with identical items or with equipment equivalent to the original in operating characteristics, mechanical strength, and ability to withstand the environmental conditions encountered in the operation of the aircraft. A suggested list of items to look for and checks to be performed are:

1. Damaged or overheated equipment, connections, wiring, and installation

2. Excessive resistance at high current carrying connections as determined by millivolt drop test
3. Misalignment of electrically driven equipment
4. Poor electrical bonding
5. Dirty equipment and connections
6. Improper support of wiring and conduit; loose connections or terminals
7. Continuity of fuses
8. Condition of electric lamps
9. Insufficient clearance or poor insulation of exposed terminals
10. Broken or missing safety wire, cotter pins, etc.
11. Operational check of electrically operated equipment such as motors, inverters, generators, batteries, lights, etc.
12. Voltage check of electrical system with portable precision voltmeter (Table 3.3)

Table 3-3.

Trouble	Probable cause	Remedy
Voltmeter registers no voltage	Voltmeter defective	Remove and replace voltmeter
	Voltmeter regulator defective	Replace regulator
	Defective exciter	Replace alternator
Low voltage	Improper regulator adjustment	Adjust voltage regulator
Erratic meter indication	Loose connections	Tighten connections
	Defective meter	Remove and replace meter
Voltage falls off after a period of operation	Voltage regulator now warmed up before adjustment	Readjust voltage regulator

13. Miscellaneous irregularities such as poorly soldered or loose swaged terminals, loose quick disconnects, broken wire bundle lacing, broken or inadequate clamps, and insufficient clearance between exposed current-carrying parts and ground

Cleaning, Preservation, Maintenance

Frequent cleaning of electrical equipment to remove dust, dirt, and grime is highly recommended. A fine emery cloth may be used to clean terminals and matting surfaces if they are corroded or dirty. Crocus cloth or very fine sandpaper may be used to polish commutators or slip rings. Do not use emery cloth on commutators because particles from the cloth may cause shorting and burning.

Accomplish adjustments to items of equipment such as regulators, generators, contactors, control devices, inverters, and relays outside the airplane on a test stand or test bench where all necessary instruments and test equipment are at hand. Follow the adjustment procedures outlined by the equipment manufacturer.

Insulation of Electrical Equipment and Bus Bar Maintenance

In some cases, a unit of electrical equipment is connected into a heavy-current circuit, perhaps as a control device or relay. Such equipment is normally insulated from the mounting structure, since grounding the frame of the equipment may result in a serious ground fault in the event of equipment internal failure. If a ground connection for a control coil must be provided, use a separate small-gauge wire.

Periodically check bus bars used in aircraft electrical systems for general condition and cleanliness. Grease, oxide, or dirt on any electrical junction may cause the connectors to overheat and eventually fail.

Clean bus bars by wiping with a clean soft cloth saturated with a standard solvent and drying with a clean soft cloth.

Replacement of Navigation and Landing Lights

Owners and pilots can change landing lights and navigation lights on their aircraft according to FAR 43. Changing a navigation light is a very easy task, comparable to changing a light bulb at home, and requires only a screwdriver. Follow these steps:

1. Ensure that the master switch is off and then remove the colored lens covering the navigation light. This is usually held in position by one screw located in the middle of the navigational light assembly (see Fig. 3-18).
2. Use a rag to cover the bulb and push down while turning the bulb counterclockwise. The rag will act as protection against cuts or lacerations should the bulb shatter.

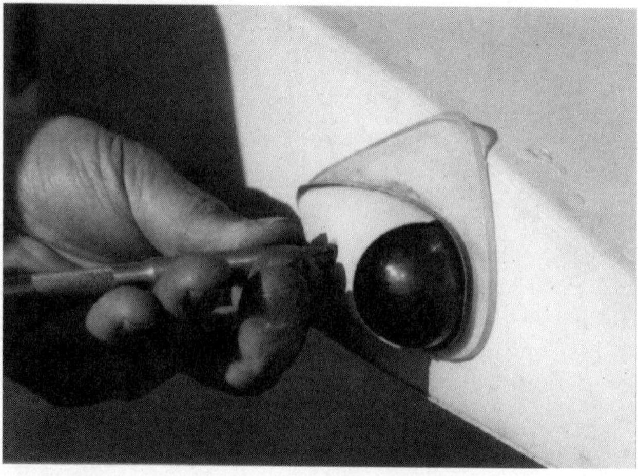

3-18 *Remove the lens covering the light bulb.*

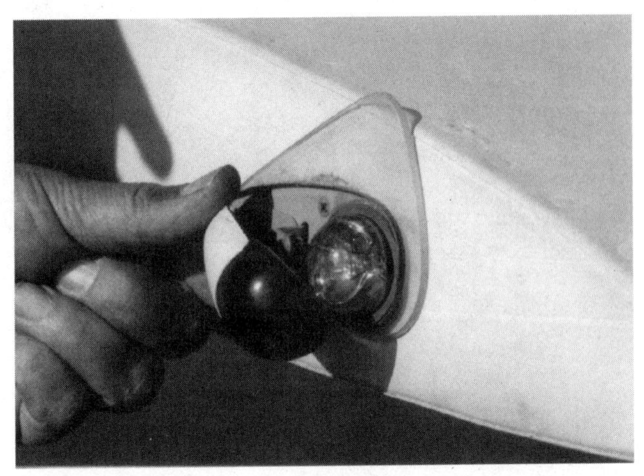

3-19 *Remove the burnt-out light bulb.*

3. Remove the bulb and discard. Insert the new light bulb in the socket and push down while turning clockwise (see Fig. 3-19).
4. Test the new bulb by turning on the navigational lights. Replace the navigation light housing.
5. Enter the procedure into the aircraft logbook. (See Chap. 8.)

Replacement of the landing light requires a screwdriver and needle nose pliers. Follow these steps:

1. Be sure the master switch is off. Rotate the propeller until it is horizontal and out of the way of the work area for the cowl-mounted landing light. Use caution whenever moving the propeller?
2. Remove the screws that hold the cover plate in position. The landing light bulb is now free to hang from its wires, so exercise caution (see Fig. 3-20).
3. Hold the bulb in one hand while unscrewing the electrical connections from the terminal at the back of the bulb.

(a)

(b)

3-20 *Remove the cover plate retaining screws: (a) cowl-mounted light; (b) wing-mounted light.*

4. Install the new bulb and connect the wires to the terminals.
5. Replace the cover plate and turn on the light for a functional check.
6. Complete the logbook entry. (See Chap. 8.)

To replace a wing-mounted landing light, follow the same procedures used for replacement of the landing light.

4

Fuel System Operation and Repair

Fuel Systems

An aircraft's fuel system is designed to provide an unrestricted flow of fuel to the engine. This flow has to be at a relatively constant pressure and must be able to meet the demands of the engine at full throttle regardless of altitude or attitude of the aircraft. In order to meet federal regulations, fuel tanks must be constructed out of metal (aluminum alloy), rubber, stainless steel, or a composite material. (See Fig. 4-1.) All metal tanks have to be able to withstand any vibration, flexing, or flight loads imposed upon them without breaking, cracking, or leaking. In addition, tanks must be able to withstand an internal pressure of 3.5 lb/in^2.

All fuel tanks must be equipped with a means of removing water or other impurities. Most tanks meeting this requirement are constructed with a sump which collects any water present in the tank, preventing the water from contaminating the fuel. In fact, 0.25 percent of the tank capacity or one-sixteenth of

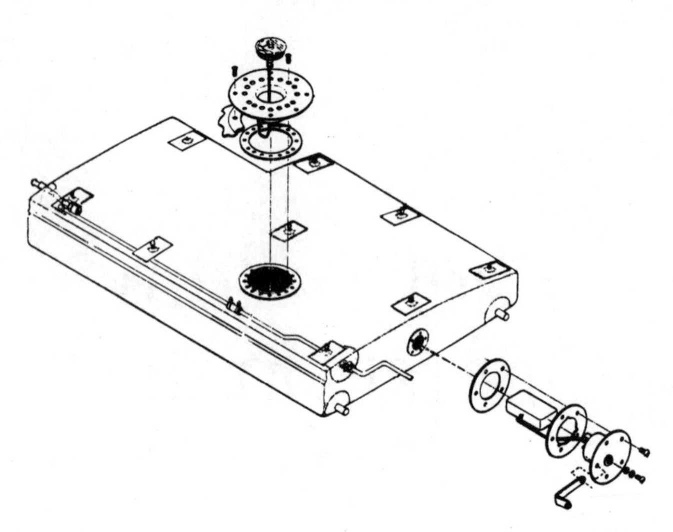

4-1 *Bladder fuel tank.*

a gallon, whichever is greater, must be set aside just for this purpose. Each sump must also have a drain which positively locks when it's closed. Some aircraft have a gascolater or sediment bowl in lieu of sumps. (See Fig. 4-2.)

Tanks must have expansion room that is at least 2 percent of the tank capacity unless the tank vent

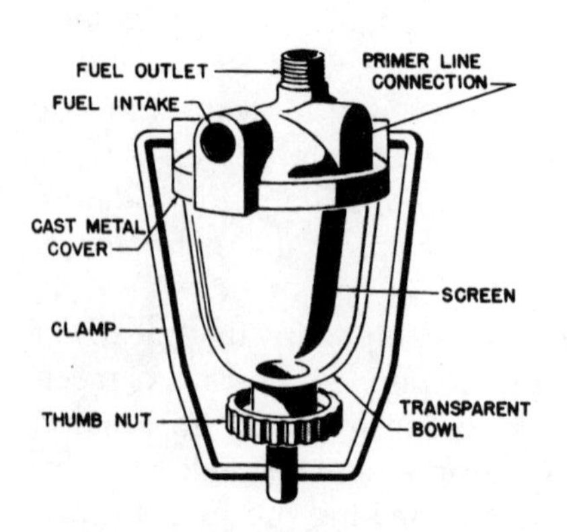

4-2 *Gascolater.*

is located in a position to overflow clear of the aircraft. This design requirement doesn't allow the expansion space to be topped off during fueling, so it's not considered part of the usable fuel load. The tank filler connection must also be designed to allow any spilled fuel to overflow down a vent tube to the ground. No fuel can spill into the fuselage or wing structure. The fuel caps may or may not be vented, but the area near the caps should be stenciled with the usable fuel capacity and the correct grade of fuel to use.

Fuel systems are designed to allow a fuel flow rate of at least 150 percent of the takeoff fuel

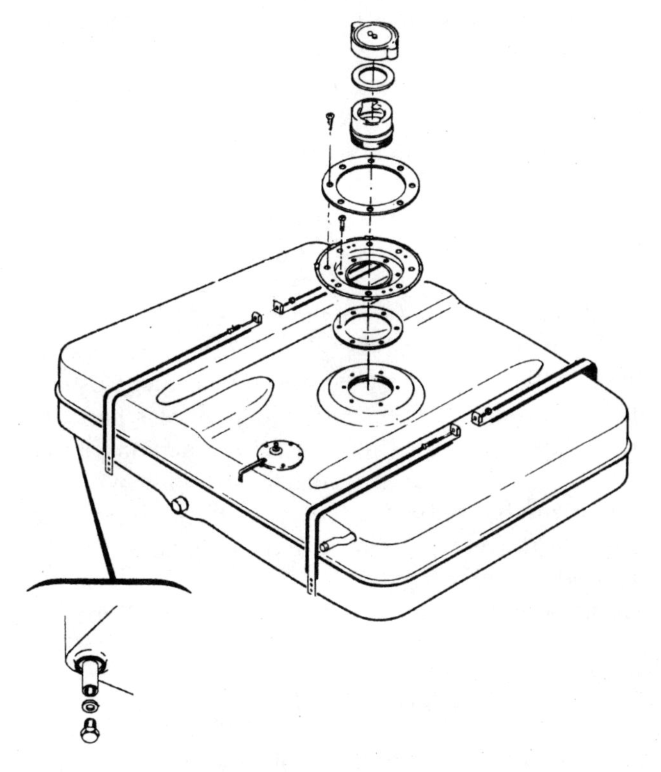

4-3 *A rigid removable fuel tank.*

requirements for a gravity-fed fuel system and 125 percent fuel flow rate of the takeoff fuel requirements for a pressure pump system. Fuel tanks come in one of three basic types: (1) the bladder, (2) a removable rigid tank, and (3) an integral tank. The bladder tank is like a rubber hot water bottle. It has all the requirements of metal tanks such as vents, sumps, and a fuel quantity system. The fuel tank is normally placed in a section of the wing or fuselage designed to withstand the weight of the fuel during normal maneuvers. The tank is usually not fuel-tight, and access is limited to a small opening. The bladder is installed through this opening and then unrolled and snapped into place. A removable rigid type of tank is found on many single-engine Cessna airplanes. (See Fig. 4-3.) The tank is an aluminum "box" welded together to form a tank. The tank is strapped into a non-fuel-tight compartment and includes vents, sumps, and a way to measure the fuel in the tank. The integral type fuel tank is built into the structure of the aircraft. Obviously, these tanks cannot be removed, since they are built into the aircraft, encompassing ribs, stringers, and spars. The tank is sealed off from the remainder of the compartment using a rubberized sealant. Some tanks have baffles built into them to prevent fuel sloshing around inside the tank. All have an access panel to allow inspection of the tank. These access panels may be on the bottom of the tank, and, obviously, the aircraft has to be defueled prior to performing an inspection. Wet wing aircraft have integral fuel tanks located in the wings. This type of tank is usually found on higher-performance aircraft and airliners.

Components

The fuel system is made up of more than just fuel tanks. Each system must have a way of determining

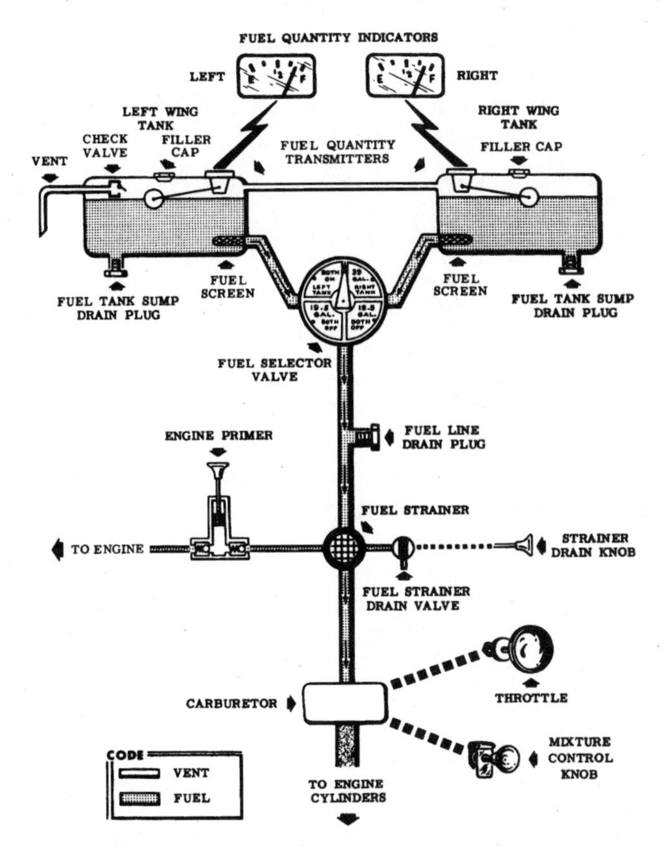

4-4 *Typical fuel system with quantity indicators.*

quantity. (See Fig. 4-4.) Some systems even indicate the fuel flow, the fuel temperature, and the fuel pressure, depending on the type of system installed in the aircraft.

Fuel Pumps

A fuel pump transfers fuel from the fuel tanks to the engine. This is accomplished by either an electric fuel pump and/or an engine-driven pump. The actual operation of the pumps is similar to any other type of pump, except, since fuel is highly volatile, the pumps have to be designed to eliminate any chance of the fuel being ignited. Each aircraft engine has a main pump, or engine-driven pump, which has the capacity to supply all normal operations. For most low-wing aircraft, an emergency fuel pump is installed as a backup to the engine-driven fuel pump. However, if the fuel system installation in the aircraft is a gravity system, an emergency pump may not be required. The fuel pump can be of the vane design or a variable volume pump. A vane pump operates by spinning a rotor, with four to six vanes, inside a housing. The vanes stay in constant contact with the sides of the housing. Fuel enters on the inlet side of the housing and is scooped up and forced out the outlet side of the housing by the rotation of the vanes. The variable-volume fuel pump (see Fig. 4-5) sends a varying amount of fuel to the carburetor depending upon the demands of the engine. Excess fuel is diverted through a pressure relief valve set at a preset value, and the fuel is routed back to the inlet side of the pump. The actual pump itself works similarly to the vane type pump.

Filters

A typical light aircraft has up to three fuel filters removing impurities from the fuel before it reaches the engine. The first filter is located at the pickup point in the fuel tank. Located at the bottom of the tank, but not in the sump, the first screen is usually a

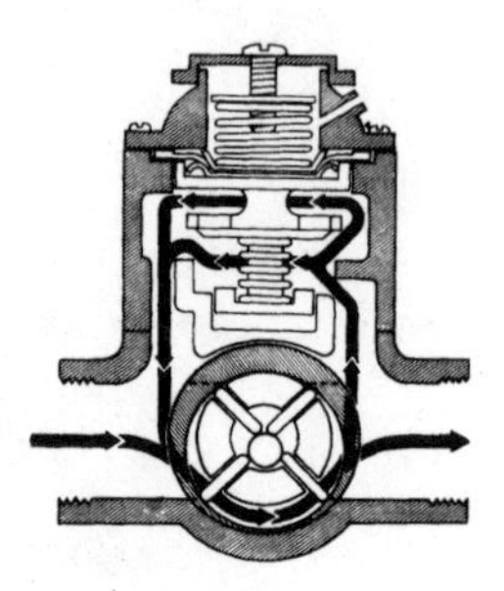

4-5
Variable-volume fuel pump.

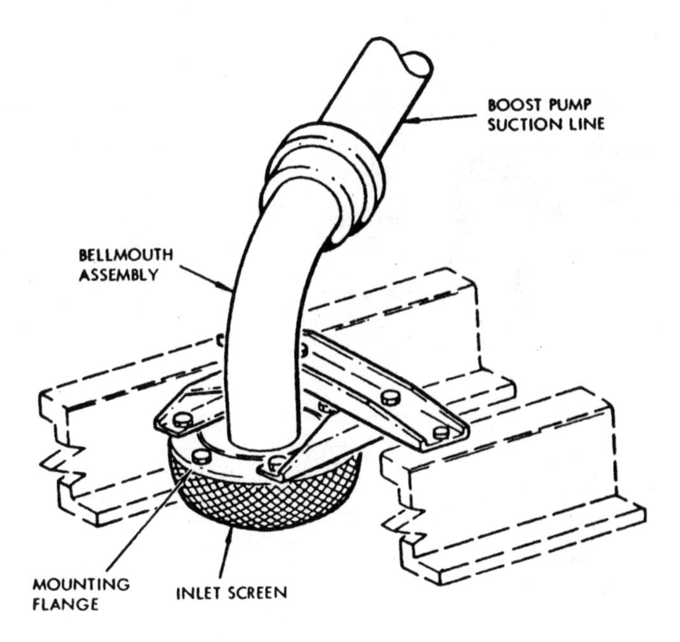

4-6 *Main fuel filter inlet screen.*

strainer that keeps larger particles of debris out of the fuel lines and system. Usually composed of a wire mesh screen, the filter has to be coarse enough to let fuel through unhindered but fine enough to stop foreign matter which could interfere with the operation of the valves and other engine components. The next filter, also known as the main fuel filter, is located at the lowest point in the fuel system. (See Fig. 4-6.) This filter is easily removable for cleaning, servicing, and inspection. The final fuel filter is located between the fuel tanks and the inlet to the carburetor or fuel injection unit. This screen is also drainable, and it too is easily removable for cleaning and inspection.

Fuel Selector and Cutoff Valves

The fuel selector valves allow the pilot to select from which tank or tanks he or she wants to burn fuel. (See Fig. 4-7.) These valves must have positive action

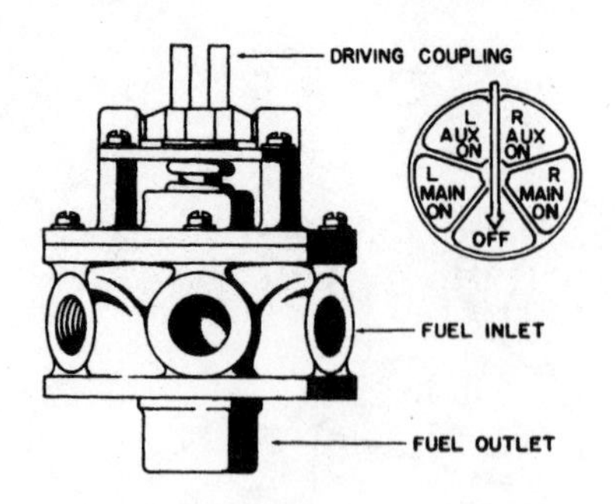

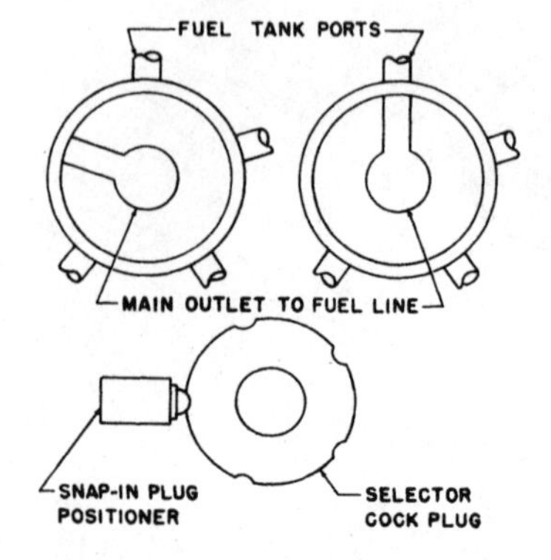

4-7 *Fuel tank selector.*

such as a detent or other means of ascertaining that the valve is in the correct position. The shutoff valve must also be positioned on the inside of the firewall and must be color-coded red. It has to be quick acting and resettable from the cockpit should the pilot inadvertently turn the fuel to the engine off.

Fuel Lines

Like hydraulic lines, fuel lines should not be allowed to sag or have sharp bends. All fuel lines should be supported with clamps to prevent vibration during engine operation. Flexible fuel lines should be incorporated in the system whenever a fuel line is attached to a structural part of the aircraft. If possible, use flex-

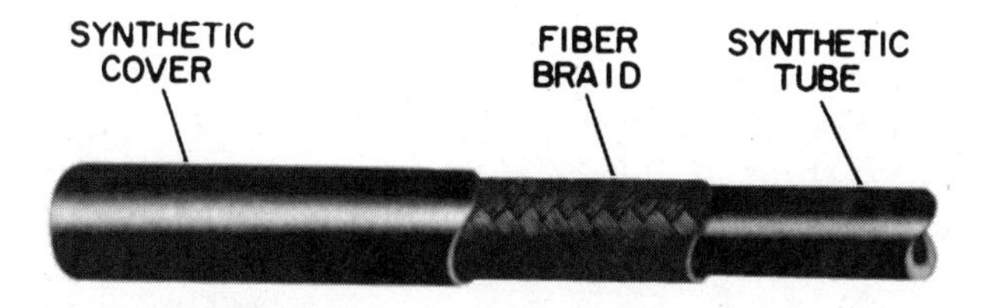

4-8 *Example of fuel hose.*

ible fuel lines whenever attaching the fuel line to the engine or any other equipment that is subject to vibrations. (See Fig. 4-8.)

Maintenance Tips

To get the best service from an aircraft's fuel system, maintain and adjust the system and its components according to the manufacturer's maintenance instructions and schedule. Some general fuel system maintenance principles are outlined below.

Fuel Lines and Fittings

Whenever fuel system lines are to be replaced or repaired, consider the following maintenance fundamentals and practices.

Compatibility of Fittings

Ensure that all fittings are compatible with their mating parts. Although various types of fittings appear to be interchangeable, they may have different threads or minor design differences which prevent proper mating and cause the joint to leak fuel or fail.

Routing

Make sure that the fuel line does not chafe against any control cables, airframe structure, etc., or come in contact with any electrical wiring or conduit. Where physical separation of the fuel lines from electrical wiring or conduit is impracticable, locate the fuel line below the wiring and clamp it securely to the airframe structure. In no case should wiring be supported by a fuel line.

Alignment

Locate all bends as accurately as possible so that the tubing is aligned with all support clamps and end fittings and is not drawn, pulled, or otherwise forced into place by them. Never install a straight length of tubing between two rigidly mounted fittings. Always incorporate at least one bend between such fittings to absorb strain caused by in-flight vibration and temperature changes.

Bonding

Bond metallic fuel lines at each point where they are clamped to the structure. Integrally bonded and cushioned line support clamps are preferred to other clamping and bonding methods.

Table 4-1. Placement of Support Clamps

If the outside diameter of the tube is	Then the approximate distance between the supports should be
1/8–3/16 in	9 in
1/4–5/16	12
3/8–1/2	16
5/8–3/4	22
1–1 1/4	30
1 1/2–2	40

Locate clamps or brackets as close to bends as possible.

Support of Line Units

To prevent possible failure, any fittings or equipment heavy enough to cause the fuel line to sag should be supported by clamps or other means rather than by the tubing itself. Place support clamps or brackets for metal fuel lines as shown in Table 4.1.

Inspection and Repair of Fuel Tanks and Cells

Welded or riveted fuel tanks that are made of commercially pure aluminum, either 3003, 5052, or similar alloys, may be repaired by welding. Tanks made from heat-treatable aluminum alloys are generally assembled by riveting. If it is necessary to rivet a new piece of aluminum in a tank, use the same type of material as used in the tank's construction and thoroughly seal the seams with a compound that is insoluble in gasoline. Special sealing compounds are available and should be used in the repair of fuel tanks. Inspect the fuel tanks and fuel cells for general condition, security of attachment, and any evidence of leakage. (See Fig. 4-9.) Examine fuel tank or cell vent line, fuel line, and sump drain attachment fittings closely.

Be sure to purge defueled tanks of explosive fuel/air mixtures in accordance with the manufacturer's service instructions. In the absence of such instructions, utilize an inert gas such as CO as a purgative to ensure the total deletion of fuel/air mixtures.

Integral Tanks

Examine the interior surfaces and seams for sealant deterioration and corrosion (especially in the sump area). Always follow the manufacturer's instructions for repair and cleaning procedures.

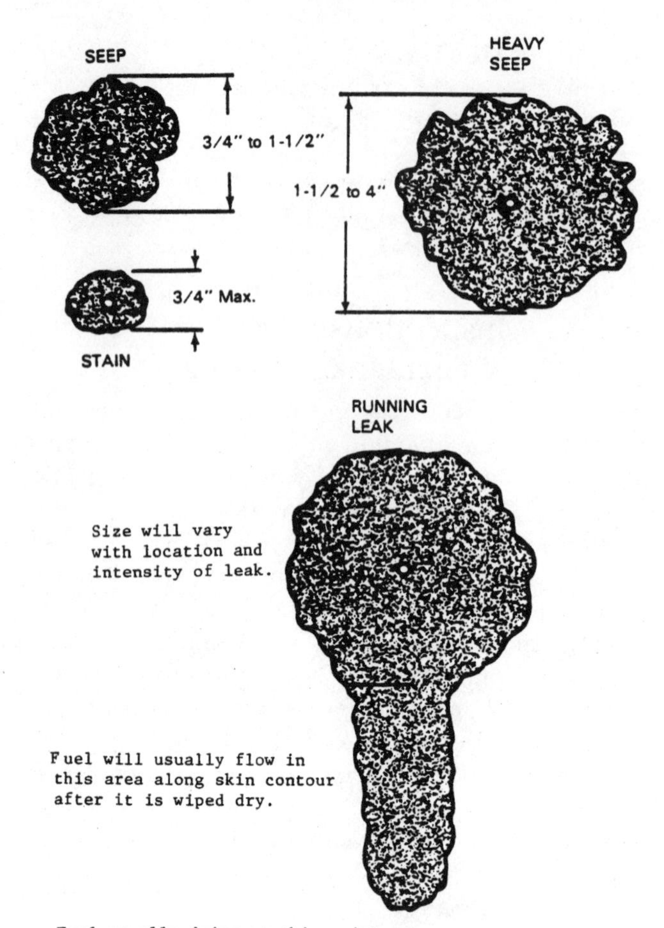

4-9 *Classification of fuel leaks.*

Internal Metal Tanks

Check the exterior for corrosion and chafing. Inspect for dents or other distortion, such as a partially collapsed tank. This collapse is usually caused by an obstructed fuel tank vent. Dents and similar damage can adversely affect accuracy of the fuel gauge and tank capacity. Thoroughly check the interior surfaces for corrosion. Pay particular attention to the sump area, especially those which are made of cast material.

Removal of Flux after Welding

It is especially important, after repair by welding, to remove completely all flux in order to avoid possible corrosion. Promptly upon completion of welding, wash the inside and outside of the tank with liberal quantities of hot water. Completely drain the tank. Next, immerse the tank in either a 5% nitric or 5% sulfuric acid solution. If the tank cannot be immersed, fill the tank with either solution and wash the outside with the same solution. Permit the acid to remain in contact with the weld about 1 hour and then rinse thoroughly with clean water. Test the efficiency of the cleaning operation by applying some acidified 5% silver nitrate solution to a small quantity of the rinse water used last to wash the tank. If a heavy white precipitate is formed, the cleaning is insufficient and the washing should be repeated.

Flexible Fuel Cells

Inspect the interior for checking, cracking, porosity, or other signs of deterioration. Make sure the cell retaining fasteners are properly positioned. If repair or further inspection is required, follow the manufacturer's instructions for cell removal, repair, and installation. Do not allow flexible fuel cells to dry out. Preserve them in accordance with the manufacturer's instructions.

Inspection of Fuel Tank Caps, Vents, and Overflow Lines

Inspect the fuel tank caps to determine that they are the correct type and size for the installation. Vented caps substituted for unvented caps may cause loss of fuel or fuel starvation. Similarly, an improperly installed cap that has a special venting arrangement can also cause malfunctions.

Unvented caps, if substituted for vented caps, will cause fuel starvation and possible collapse of the fuel tank or cell. Malfunctioning of this type occurs when the pressure within the tank decreases as the fuel is withdrawn. Eventually, a point is reached where the fuel will no longer flow and/or the outside atmospheric pressure collapses the tank. Thus, the effects will occur sooner with a full fuel tank than with one partially filled.

Check tank vents and overflow lines thoroughly for condition, obstructions, correct installation, and proper operation of any check valves and ice protection units. Pay particular attention to the location of the tank vents when such information is provided in the manufacturer's service instructions. Inspect for cracked or deteriorated filler opening recess drains, which may allow spilled fuel to accumulate within the wing or fuselage. One method of inspection is to plug the fuel line at the outlet and observe fuel placed in the filler opening recess. If drainage takes place, investigate the condition of the line and purge any excess fuel from the wing.

Ensure that filler opening markings are stated according to the applicable airworthiness requirements and are complete and legible.

Inspection and Repair of Fuel Cross-Feed, Firewall Shutoff, and Tank Selector Valves

Inspect these valves for leakage and proper operation as follows.

Internal and External Leakage

Internal leakage can be checked by placing the appropriate valve in the off position, draining the fuel strainer bowl, and observing if fuel continues to flow into it. Check all valves located downstream of boost pumps with the pumps operating. Do not operate the pumps longer than necessary. External leakage from these units can be a severe fire hazard, especially if the unit is located under the cabin floor or within a similarly confined area. Correct the cause of any fuel stains associated with fuel leakage.

Selector Handles

Check the operation of each handle or control to see that it indicates the actual position of the selector valve. Make sure that stops and detents have positive action and feel. Worn or missing detents and stops can cause unreliable positioning of the fuel selector valve.

Worn Linkage

Inaccurate positioning of fuel selector valves can also be caused by worn mechanical linkage between the selector handle and the valve unit. An improper fuel valve position setting can seriously reduce engine power by restricting the available fuel flow. Check universal joints, pins, gears, splines, cams, levers, etc., for wear and excessive clearance, which prevent the valve from positioning accurately or from obtaining fully off and on positions.

Ensure that all required placards are complete and legible. Replace those that are missing or cannot be read easily.

Fuel Pumps; Fuel Filters, Strainers, and Drains; Indicator Systems

Inspect, repair, and overhaul boost pumps, emergency pumps, auxiliary pumps, and engine-driven pumps in accordance with the appropriate manufacturer's instructions.

Check each strainer and filter element for contamination. Determine and correct the source of any contaminants found. Replace throwaway filter elements with the recommended type. Examine fuel strainer bowls to see that they are properly installed according to direction of fuel flow. Check the operation of all drain devices to see that they operate properly and have positive shutoff action.

Inspect, service, and adjust the fuel indicator systems according to the manufacturer's instructions. Determine that the required placards and instrument markings are complete and legible.

Owner Maintenance: Fuel Line Replacement

A fuel leak or chaffing of a fuel line may be reason to change a prefabricated fuel line. It requires a few hand tools, and care must be exercised while working around gasoline. Be sure to ground the aircraft and use plastic containers to catch any leaking fuel while removing the old fuel line. Tools needed are an open-ended wrench of the appropriate size, a screwdriver, and safety wire and pliers. Follow these steps:

1. Turn off the fuel shutoff valve. Remove the cowling to gain access to the damaged fuel line.
2. Using an open-ended wrench loosen the connection between the fuel line and the connection point. Be careful when removing the line, as residual fuel may spill out. Be sure a plastic container is underneath the work area.
3. Inspect the new fuel line for any obvious damage. Compare the new line to the fuel line that has been removed for correct size

and shape. Install the fuel line and tighten the connections as called for in the maintenance manual and safety wire. Be sure that the fuel line does not rest on any other surfaces or will chafe against any control lines or wires.

4. Be sure the aircraft is securely chocked and turn the fuel shutoff valve back to on. Run the engine for several minutes, then shut down and check for any leaks.
5. Replace cowling if no leaks were noted.
6. Complete logbook entry. (See Chap. 8.)

5

Repair of Hydraulic Systems

Including Wheels and Brakes

Hydraulic Systems

Light-aircraft hydraulic systems operate by utilizing a power section and an actuating section. The power section is responsible for moving the hydraulic fluid under pressure through the hydraulic lines. (See Fig. 5-1.) The actuating system operates the system components, such as the landing gear, the flaps, brakes, or any other hydraulic unit. The power section is pressurized from either an engine-driven hydraulic pump or an electric hydraulic pump. The pump moves the hydraulic fluid through a hydraulic system which is classified as either open or closed. (See Fig. 5-2.) The open hydraulic system has hydraulic fluid continuously circulating through the

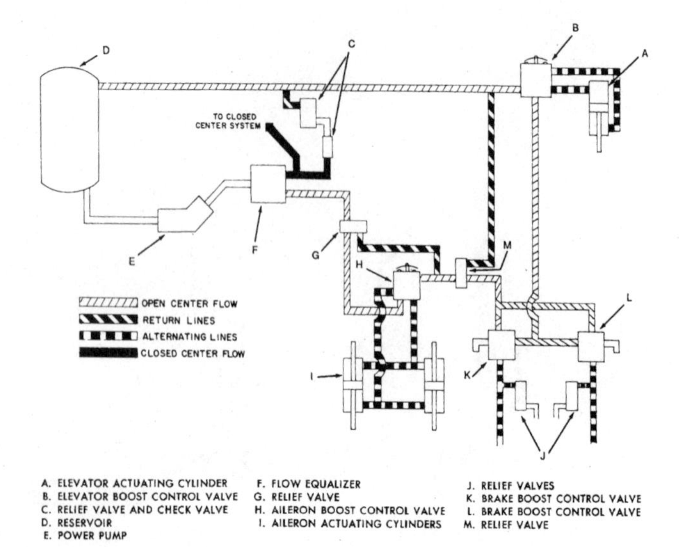

5-1 *Power section with open-center flow. (Courtesy of Michael J. Kroes, William A. Watkins, and Frank Delp, Aircraft Maintenance and Repair, 6th ed., McGraw-Hill, 1993)*

lines and units such as gear flaps are actuated by bypassing the moving fluid into the unit being operated. Once the desired results are achieved, i.e., the gear retracts, the flaps move, and so on, then the system reverts to a free-flowing system. The other type of hydraulic system is a closed system. This hydraulic system always has pressure built up to a predetermined value by either an engine-driven or electric hydraulic pump. A pressure-regulating valve unloads the pressure once it builds up to a preset level. In this way, the pump is allowed to idle when no demands are placed upon the system.

The most common system used on light aircraft is the open system because of its lack of complexity. However, a major disadvantage to an open system is that during operation of more than one hydraulic component at a time, such as simultaneously operating the gear and flaps, the component that is the first unit downstream of the pump will operate at a normal rate while the second component will operate at a much slower rate. This slower rate will continue until the first system completes its cycle and the selector valve closes to the first component. Once this occurs, the second system will operate normally. Many aircraft manufacturers have eliminated this problem by utilizing a hydraulic "powerpack." A powerpack is the hydraulic reservoir, relief valves, emergency hand pump, and other components, all assembled in one

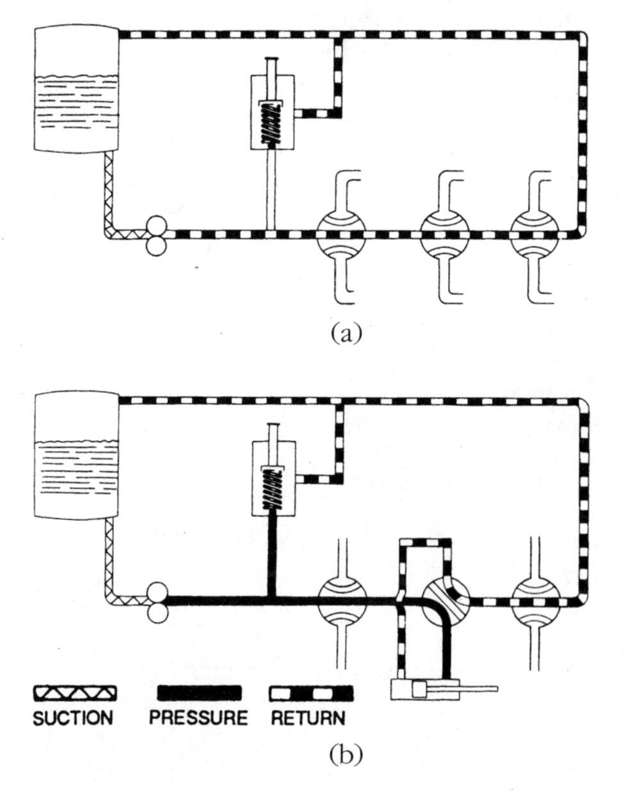

5-2 *(a) Closed system. (b) Open system. (Courtesy of Michael J. Kroes, William A. Watkins, and Frank Delp, Aircraft Maintenance and Repair, 6th ed., McGraw-Hill, 1993)*

modular type unit. The powerpack works on the principle of an open system in that the fluid circulates under a reduced pressure until a component is actuated. A valve then moves to direct the hydraulic fluid to that unit and locks out any other component from operating. Once the desired results are achieved, the hydraulic pressure builds up and the valve returns to an open position, allowing the fluid to again circulate.

The hydraulic systems of most light aircraft are relatively simple and easy to repair. Most hydraulic components, including pumps, actuating cylinders, and valves, should be repaired or replaced according to the manufacturer's instructions. Some general principles of maintenance and repair are outlined below.

Inspection and Repair of Components

Hydraulic components such as pumps, actuating cylinders, selector valves, relief valves, should be repaired or adjusted following the airplane and component manufacturer's instructions. Inspect hydraulic filter elements at frequent intervals and replace as necessary.

Cleaning and Lubricating

It is recommended that only easily removable neutral solutions be used when cleaning landing gear components. Any advantage, such as speed or effectiveness, gained by using cleaners containing corrosive materials can be quickly counteracted if these materials become trapped in close-fitting surfaces and crevices. Wear points, such as landing gear up-and-down latches, jackscrews, door hinges, pulleys, cables, bell-cranks, and all pressure-type grease fittings should be relubricated after every cleaning operation. To obtain proper lubrication of the main support bushings, it may be necessary to jack the aircraft. *Note:* Any time the aircraft is on jacks, check the landing gear main support bushings for wear. Consult the aircraft manufacturer's overhaul manual for specific wear tolerances.

During winter operation, excess grease may congeal and cause increased loads on gear retraction system electric motors and hydraulic pumps. This condition can lead to component malfunctions, so be sure to clean off any excess grease during and after lubrication.

Emergency Systems

Exercise the emergency landing gear systems periodically to ensure proper operation and to prevent inactivity, dirt, and corrosion from rendering the system inoperative. Most emergency systems employ a mechanical, pressure-bottle, or free-fall extension capability. Check that triggering mechanisms are in working order and for the presence of required placards. Be sure that all necessary accessories such as cranks, levers, handles are readily available.

Special Inspections

Any time an aircraft has experienced a hard or overweight landing, a special structural inspection, which includes the landing gear, should be performed. Typical areas which require special attention include the following: landing gear support trusses for cracked welds, sheared bolts and rivets, and buckled structures; wheels and tires for cracks and cuts; and upper and lower wing surfaces for wrinkles, deformation, and loose or sheared rivets.

Retraction Tests

Periodically, perform a complete operational check of the landing gear retraction system. Inspect the normal extension and retraction system, the emergency extension system, and the indicating and emergency warning system. Determine that the actuating cylinders, linkage, slide tubes, sprockets, chain or drive gears, gear doors, and the up-and-down locks are in good condition and properly adjusted and lubricated. In addition, an electrical continuity check of microswitches and associated wiring is recommended. Only qualified personnel should attempt adjustments to the gear position and warning system microswitches, and then only by closely following the manufacturer's recommendations.

Wheels

Inspect the wheels at periodic intervals for any cracks, corrosion, dents, distortion, or faulty bearings in accordance with the manufacturer's service bulletins. (See Fig. 5-3.) In split-type wheels, recondition bolt holes which have become elongated by the use of inserts. Pay particular attention to the condition of the through-bolts and nuts. Carefully inspect the wheels used with tubeless tires for damage to the wheel flange and for proper sealing of the valve. The sealing ring used between the wheel halves should be free of damage and deformation. In bolting wheel halves together, be sure to tighten the nuts to the proper torque value. Periodically accomplish an inspection to make sure the nuts are tight and there is no movement between the two halves of the wheel. Maintain the grease-retaining felts in the wheel assembly in a soft, absorbent condition. If any have become hardened, wash them with a petroleum-base cleaning agent; if this fails to soften them, they should be replaced.

Corrosion of Wheels

Thoroughly clean the wheels if corroded and then examine for soundness. Smooth and repaint bare

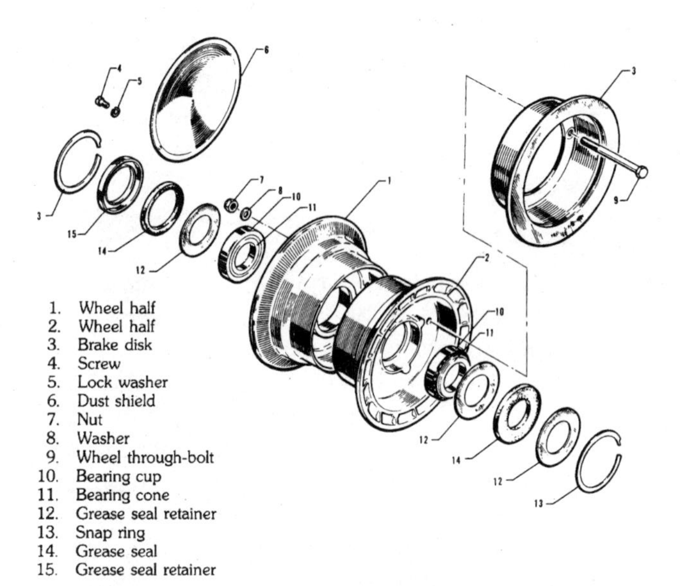

5-3 *Example of a main wheel assembly. (Courtesy of Piper Aircraft, Inc.)*

corroded spots with a protective coating such as zinc chromate primer and aluminum lacquer or some other equally effective coating to prevent further corrosion. It is necessary to always replace wheels having severe corrosion which might affect their strength.

Dented or Distorted Wheels

Replace any wheels that wobble excessively due to deformation resulting from a severe side-load impact. Dents of a minor nature do not affect the serviceability of a wheel.

Wheel Bearings

Periodically inspect wheel bearings for condition. Replace damaged or excessively worn parts. Maintain bearings and races as matched sets. Repack bearings with a high-melting-point grease prior to their installation. Avoid preloading the wheel bearing when installing on aircraft by tightening the axle nut just enough to prevent wheel drag or side play.

Brakes

Maintain the clearance between the moving and stationary parts of a brake in accordance with the manufacturer's instructions. Disassemble and inspect the brake periodically and examine the parts for wear, cracks, warpage, corrosion, elongated holes, and so on. Discolored brake disks are an indication of overheated brakes and should be replaced. If any of these or other faults are indicated, repair, recondition, or replace the affected parts in accordance with the manufacturer's recommendations. Surface cracks on the friction surfaces of the brake drums occur frequently owing to high surface temperatures. These surface cracks may be disregarded as seriously affecting the airworthiness until they become cracks of approximately 1 in in length.

Hydraulic Brakes

For proper maintenance, periodically inspect the entire hydraulic system from the reservoir to the brakes. Maintain the fluid at the recommended level with proper brake fluid. (See Fig. 5-4.) When air is present in the brake system, bleed it out in accordance with the manufacturer's instructions. Replace all flexible hydraulic hoses that have deteriorated because of long periods of service. Replace hydraulic piston seals when they show evidence of leakage.

Microswitches

Inspect microswitches for security of attachment, cleanliness, general condition, and proper operation. (See Fig. 5-5.) Check associated wiring for chafing and proper routing and to determine that protective covers are installed on the wiring terminals. Check the condition of the rubber dust boots which protect the microswitch plungers from dirt and corrosion.

Hydraulic Lines and Fittings

When inspecting hydraulic lines for airworthiness, check for any evidence of leaking fluid, especially around fittings or connections. Inspect for any kinks or pin holes in hydraulic lines and ensure all connections are tight. If a metal line is discovered to be damaged or defective, replace the entire line. Be sure to use the same size hydraulic tubing and material as the line being replaced. If it's not too damaged, the old hydraulic line makes an excellent template for the new hydraulic line.

If the hydraulic tube has to be bent, ¼-in soft aluminum can be bent by hand, but any greater diameter should be bent by a tube-bending tool. When bending tubing, attempt to avoid any flattening in the bends. During installation, ensure that the hydraulic tube that you replaced lines up correctly with the original tube or part. Do not force the hydraulic tube into alignment using the coupling nuts as a wedge. Some connections will require flaring the tubing's end. Always use the correct tool for this important job. Aviation flare-forming tools flare the tubing at 37 degrees. The automotive flare tool flares at 45 degrees and should not be used.

When making the tubing connections, here's a word of caution: Besides flaring, a common connection method is to employ standard connection fittings such as an AN-818 nut and an AN-819 sleeve. Do not overtighten these, since damage to the flare, the tubing, or the fittings could result. Use hydraulic fluid as lubricant while tightening the connection to provide an assortment of tube data.

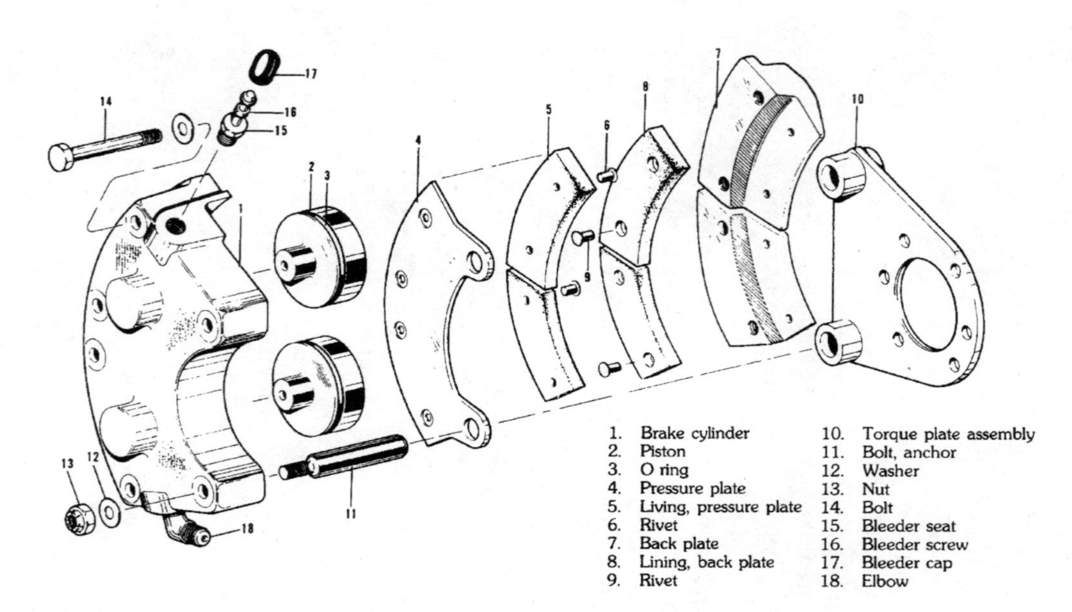

5-4 *Brake components. (Courtesy of Piper Aircraft Inc.)*

5-5 *Example of a microswitch. (Courtesy of Honeywell's Microswitch Division)*

Some metal tubes may not need replacing but can be repaired instead. Minor dents, scratches, and nicks can usually be repaired by burnishing out the damage if the nicks are no deeper than 10 percent of the wall thickness. Any dents or nicks that are situated in the heel of a bend are reason enough to replace the tubing.

Follow the same guidelines above when replacing flexible hydraulic lines. Always use the same size, type, and length of hose when replacement of a flexible line is required. Allow for about 5 to 8 percent slack in the hydraulic hose between the two fittings. If the hose has no play, it will become overstressed and eventually fail.

Teflon hoses are superior to rubber hoses in high-temperature locations. Consider using a Teflon hose in the engine compartment or any other location where heat may be a concern. Remember Teflon hoses require a bit of extra care when handling: Do not allow them to kink by excessive bending or twisting. Kinks will eventually cause weakness in the tubing wall. Also be aware that if Teflon tubing is already in service in the aircraft, do not attempt to straighten a hose that has been removed from the aircraft. Once in use, Teflon tends to assume a rather permanent set, especially if it's been exposed to high temperatures or pressures.

O-Rings

A thorough understanding of the O-ring seal's uses and applications is necessary to determine when replacement of the seals should be made. (See Fig. 5-6.) The simplest application of an O-ring is where it merely serves as a gasket usually being compressed within a recessed area by applications of pressure with a packing nut or screw cap. Leakage is not normally acceptable in this type of installation. In other installations, the O-ring seals depend primarily upon their resiliency to provide sealing action. When moving parts are involved, minor seepage may be normal and acceptable. A moist surface found on moving parts of hydraulic units is an indication the seal is being properly lubricated. In pneumatic systems, seal lubrication is provided by the installation of a grease-impregnated felt wiper ring. When systems are static, seepage past the seals is not normally acceptable.

Inspection of O-Ring Seals

During inspection, consider the following to determine whether O-ring seal replacement is necessary:

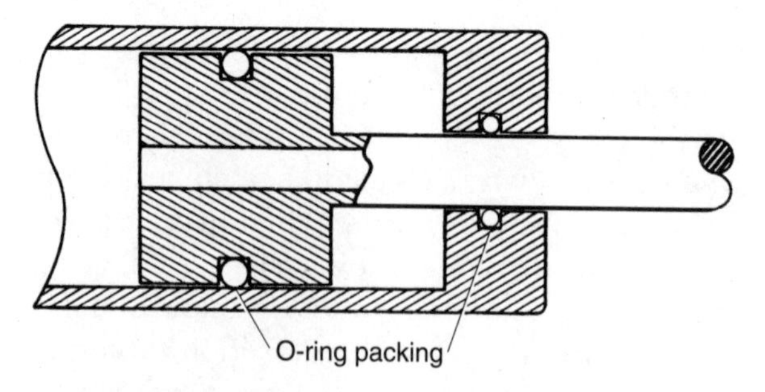

5-6 *Typical O-ring installation. (Courtesy of Michael J. Kroes, William A. Watkins, and Frank Delp, Aircraft Maintenance and Repair, 6th ed., McGraw-Hill, 1993)*

How much fluid or air is permitted to seep past the seals? In some installations minor seepage is normal. Always refer to the manufacturer's maintenance manuals for more information.

Does the leak of fluid or air create a hazard or affect surrounding installations? A check of the system fluid and a knowledge of previous fluid replenishment is helpful.

Will the system function safely until the next inspection?

O-Ring Seal Maintenance

Following are tips for good O-ring maintenance:

Correct all leaks from static seal installations. Avoid using tools that might damage the seal or the sealing surfaces.

Be sure that the part number is correct.

Retain replacement seals in their package until ready to use. This provides proper identification and protects the seal from damage and contamination.

Make sure that the sealing surfaces are clean and free of nicks or scratches before installing seal.

Protect the seal from any sharp surfaces that it must pass over during installation. Use an installation bullet or cover the sharp surfaces with tape.

Lubricate the seal so it will slide into place smoothly.

Allow sufficient time for the seal to cold-flow to its original size before continuing with the installation.

To avoid maintenance mistakes:

- Don't retighten packing gland nuts; retightening will, in most cases, increase rather than decrease the leak.
- Never reuse O-ring seals because they tend to swell from exposure to fluids and become set from being under pressure. They may have minor cuts or abrasions that are not readily discernible by visual inspection.
- Do not depend upon color coding. Coding may vary with manufacturer.
- Don't allow the seal to become twisted.

Storage of Replacement Seals

Store O-ring seals where temperatures do not exceed 120°F, and keep seals packaged to avoid exposure to ambient air and light, particularly sunlight. Avoid storing O-ring seals on pegs or hooks, as in hardware stores; this practice leads to loss of identification and loss of cure dates, impairs cleanliness, and tends to cause the seals to become distorted and lose their resiliency from exposure to light and ambient air.

Adding Hydraulic Fluid

Adding hydraulic fluid to the aircraft system is a simple process. Refer to the aircraft maintenance manual to determine the correct type of hydraulic fluid to use. Do not mix fluid types or use a type other than the one recommended. Hydraulic fluids are not all interchangeable. Most light aircraft utilize MIL-H-5606 mineral-based hydraulic fluid, which is red in color. Avoid skin contact with the fluid while replenishing the reservoir. You will need a screwdriver, funnel, and flashlight; follow these steps:

1. Gain access to the hydraulic reservoir by removing the cowling or access plate. Open

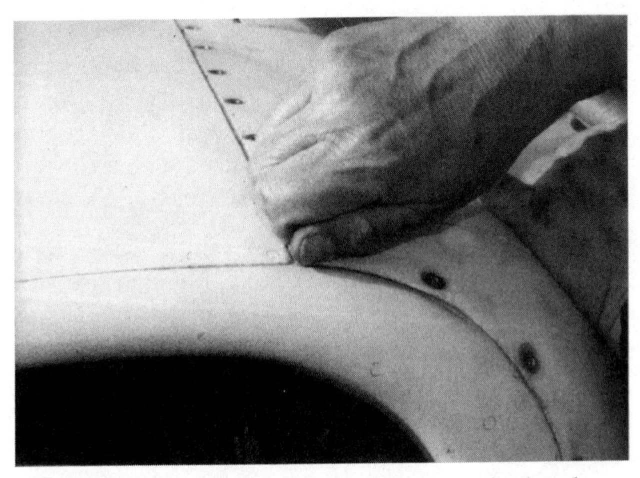

5-7 *Remove the cowling to gain access to hydraulic reservoir.*

the filler cap on the reservoir and use the flashlight to determine approximately how much fluid needs to be added to bring the fluid up to the full mark on the reservoir (see Fig. 5-7).

2. Use the funnel and pour in the approximate amount of fluid. Check to see if the reservoir is full. Repeat this process until full.
3. Replace the filler cap and clean up any spilled hydraulic fluid immediately.
4. Replace the cowling or close the access panels and test the system.
5. Make the appropriate logbook entry. (See Chap. 8.)

Wheel Bearings

Wheel bearings require periodic inspections and repacking with grease. An excellent time to accomplish this task is while the wheel has been removed to change a tire. All that are required to inspect and repack wheel bearings are a few simple hand tools assuming the aircraft wheel has already been removed. A magnifying glass helps to inspect the bearings, and a soft toothbrush is used with a solvent to clean the bearings. Follow these steps:

1. Remove the retaining clip from the wheel to access the bearings. Most light aircraft use roller bearings set at a slight angle or taper (Fig. 5-8).
2. Use a clean rag to remove the majority of old grease from the bearings. Use Varsol or some other approved commercial solvent to clean the remainder of the grease from the bearings. This is where the toothbrush is helpful.
3. Once clean, inspect the bearings for any damage or corrosion. A magnifying glass and a strong light are very helpful in this process. Look for any cracks or damage to the bearings or the race. If damage is present, replace the wheel bearing with an approved part.

5-8 *Remove retaining bolt to gain access to wheel bearings.*

4. Repack the bearing with an approved grease. Be sure to get the grease between all the roller bearings. The best method is to place a small amount of grease in the palm of the hand and press the bearing into it until grease is coming out the other side of the bearing.
5. Wipe away the excess grease being careful not to remove the grease from the bearing surfaces.
6. Replace the bearing in the wheel and secure.
7. Complete the logbook entry. (See Chap. 8.)

6

Tire Wear, Repair, and Replacement

Tires

Proper inflation is very important to the condition and longevity of aircraft tires. The manufacturer's inflation pressure is for an unloaded tire, that is, a tire that is off the aircraft and has no load on it. When inflating the tire while it is on the aircraft, adjust the pressure about 5 percent higher than an unloaded tire. This compensates for the decreased volume caused by the weight of the aircraft bulging the tire.

An airplane's tires operate in an environment more severe than stock car racing. Aircraft tires have to endure not only high speeds but also heavy loads, especially during a hard landing. In addition, aircraft tires are manufactured using rubber, a material which is a very good insulator; because it's a good insulator, all the heat generated from fast taxiing and heavy braking takes a long time to dissipate. Improper inflation coupled with this residual heat significantly deplete a tire's useful life.

Proper inflation is important for another reason. Tires are manufactured with layer upon layer of fabric overlapping each other. Uninflated and off the aircraft, each layer of fabric is designed to support the loads placed on the tire in a uniform manner. Normal loading and impact forces from landings (and bounces!) results in a higher shear load placed on the outside plies of the tire than on the inside plies. This shearing motion is exacerbated when the tire is either underinflated or overinflated. Check the tire's air pressure at least once a week if not during each preflight. (See Fig. 6-1.) Try to use a dial gauge rather than a plunger type design. (See Fig. 6-2.) Dial gauges are more expensive, but they are much more accurate.

Choosing a tire depends on the weight and size of the aircraft and the normal landing speed. Although the speeds and loads imposed on an aircraft tire are much greater than the family automobile, the tires are manufactured in much the same way, with the exception of using a greater number of fabric plies. A ply rating is only an indication of a

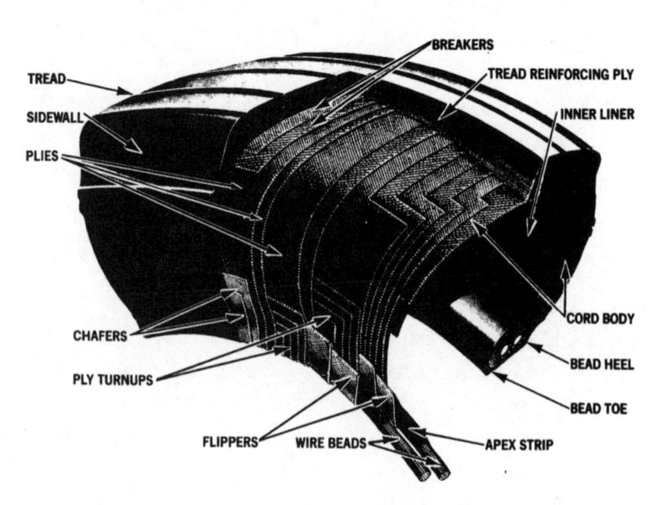

6-1 *Parts of a tire. (Courtesy of Goodyear Aviation Products)*

6-2
Example of a dial gauge. (Courtesy of Boeing Commercial Aircraft Co.)

tire's strength. The higher the ply rating, the greater a load the tire can safely support. A common misconception among pilots, and even some mechanics, is that a ply rating represents the number of fabric layers in a tire. That's not necessarily true, rather, the ply rating is used to determine the tire's strength and possible usage. When changing an aircraft tire, note that tubes and tires have a balance mark that should be lined up to help balance the tire. These marks, usually a red dot, are placed on the lightest part of the tire by the manufacturer. Tires in use today are classified as follows:

III—a low-pressure, low-speed tire

VII—very high pressure, low- or high-speed design

VIII—very high pressure, low- or high-speed with a low profile.

The tires are manufactured under a technical standard order (TSO). When purchasing a tire, check the side wall to determine if the tire meets the provisions of FAR 37.167. Each tire will be permanently stamped with the name of the manufacturer, the country where manufactured, the serial number, and the ply rating. (See Fig. 6-3.) Also included is the tire size and the TSO number. Tires designed to operate at speeds greater than 160 mph will also have a speed rating and a skid depth printed on them. Of course,

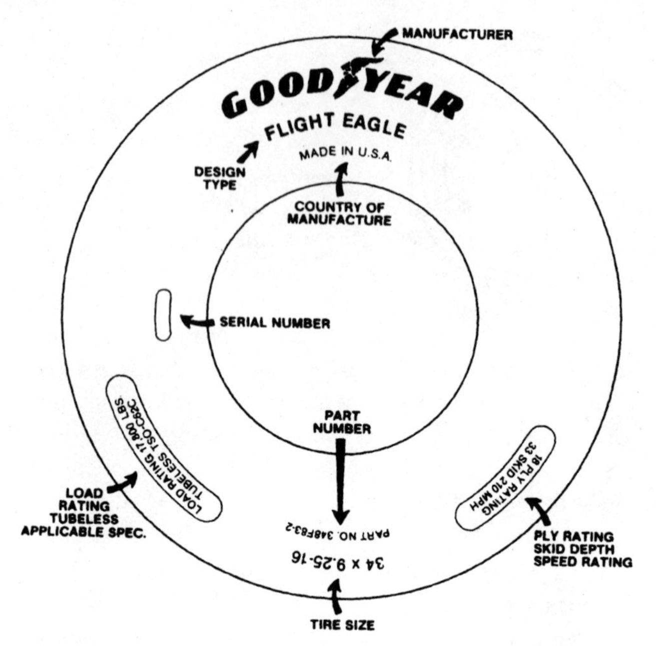

6-3 *Tire markings. (Courtesy of Goodyear Aviation Products)*

as the economy goes global, and metric becomes the aviation standard, the tire industry is also joining this trend. Some new tire classifications are already in use. These classifications include a three-part numeric size designation. The new tires will be classified as follows:

> *Outside diameter.* If the metric system is used, expect to see a millimeter size here.
>
> *Width.* This is the rim width and is listed as a percentage.
>
> *Bead diameter.* This is the angle of the base of the tire bead.

Tires are designed and built to withstand a lot of punishment. Accelerating from zero to over 100 mph on landing, heavy use of brakes, and the side loads imposed by a fast turnoff all take their toll on aircraft tires. The manufacturers have built extremely tough and durable tires engineered to take this repeated punishment. During construction, steel wire beads are used to form the foundation of a tire. Next, the plies are added. These are diagonally placed layers of nylon cords where each layer is laid in an alternating position to the previous layer. This adds strength to the tire. The beads are made of reinforced rubber and have steel wires running through them. This provides a firm mounting area for attaching the tire to the wheel. The tread is the part of the tire that comes into contact with the ground. The grooves are designed to provide maximum traction under a variety of runway and weather conditions. (See Fig. 6-4.)

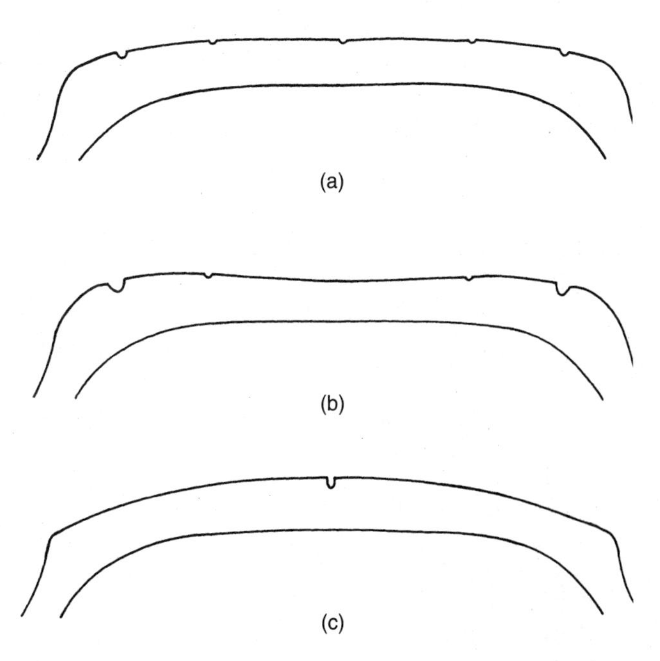

6-4 *Examples of tread wear indicating (a) normal inflation, (b) overinflation, (c) underinflation.*

Inspection and Repair of Aircraft Tires

When storing aircraft tires, always place them on edge to prevent them from distorting. If they are placed in a stack, try to limit the stack to only three tires. Try to avoid placing tires in sunlight or near electrical equipment emitting ozone. Ozone attacks rubber and causes it to prematurely age. Of course, don't allow any chemicals or fuel oils to come in contact with stored tires. It is essential that tires be inspected frequently for cuts, worn spots, bulges on the sidewalls, foreign bodies in the treads, and tread condition. Some defective or worn tires may be repaired or retreaded in accordance with manufacturer's standards. The term *retread* refers to the several means of restoring a used tire, whether by applying a new tread alone or tread and sidewall material in varying amounts. The term refers as well to the process of extending new sidewall material to cover the bead area of the tire. Repairs are also included in the retreading of tires.

Repair and Retreading of Low-Speed Tires

The following procedures are applicable to other than type VII and type VIII high speed aircraft tires. Tires with injuries of the following types may be repaired:

Bead injuries where only the chafe-resistant material is damaged or loose, or where minor injuries do not penetrate into more than 25 percent of the tire plies, up to a maximum of three damaged plies

Injuries in tread or sidewalls may be repaired by the spot repair method; this includes cuts in the tread area that are smaller than ½ in in length and

Table 6-1. Cut Depth in Tire Treads

Ply rating	Maximum cut depth in plies
<8	0
8–16	2
16+	4

do not penetrate more than a certain number of plies into the cord body (see Table 6-1).

If any of the following conditions exist, repair of the tire is not recommended:

Evidence of flex breaks

Bead injuries affecting the seal of the bead on tubeless tires

Evidence of separation between plies or around bead wire

Injuries requiring reinforcement

Kinked or broken beads

Weathering or radial cracks extending into the cord body

Evidence of blisters or heat damage

Cracked, deteriorated, or damaged inner liners of tubeless tires

Spot Repairs

Use repair methods conforming to the best aviation industry practices. Skive injuries require spot repairs at an approximate 45-degree angle to remove damaged rubber and cords.

Buff away the exposed cord leaving only the ends at the skive line and the rubber between the exposed cord plies. Clean the area thoroughly before applying the vulcanizing cement. As soon as the cement is completely dry, cover the skived area of the carcass with repair cushion gum. Fill the cavity in the carcass with repair cushion gum and roll to remove air. The tread cavity can then be filled with tread repair gum well-rolled, to provide a solid repair.

Bead or Chafe Repairs

Trim the frayed fabric ends. Turn back the loose area, buff and clean the surface, and immediately cement back into place. Always vulcanize repairs larger than 2 in.

Owner Maintenance—Changing Aircraft Tires

Changing an aircraft tire is considered preventative maintenance by the FAA and therefore can be accomplished by the aircraft owner. Changing a tire is not a difficult task, but it does require using the correct tools and having the knowledge to perform the task. The best source of information is in the aircraft's maintenance manual. but the following information can be utilized as a reference.

You will need an aircraft jack, a socket wrench of the correct size, and a ratchet wrench. Follow these steps:

1. Jack the aircraft and remove the wheel from the axle. Be careful to protect the bearings from exposure to dirt or damage.
2. Once the tire is removed, deflate it by removing the core of the air valve.
3. Lay the wheel on a flat surface or workbench and break the tire beads away from the

wheel's rim. Try to apply an even pressure around the sidewalls of the tires. Work as close to the tire beads as possible. Do not use a screwdriver or other sharp tool to pry the tire away from the wheel. If possible, use a demounting machine at a local shop.

4. Now remove the nuts and bolts holding the two halves of the wheel together. Remove the wheel halves from the tire.
5. Inspect the tube and determine if it is airworthy. If in doubt, replace it with a new tube.
6. Inspect the new tire for any manufacturing defects. Dust the tube with tire talc. This will help prevent the tube from sticking to the tire. Insert the tube into the tire. Next, align the yellow mark on the tube with the red dot on the tire. Inflate the tube slightly, just enough to give it shape and remove any wrinkles. Allow it to deflate.
7. Replace the wheel half containing the valve stem first. Pull the valve stem into position, then install the second half of the wheel.
8. Install the wheel bolts and tighten with a torque wrench in alternating sequence.
9. Reinstall the valve core and inflate the tire to normal pressure using shop air. Some installations call for nitrogen, but most light airplanes use air. Be careful during the inflation process to prevent the tire from rupturing. If possible, inflate the tire in an inflation booth or other containment area.
10. Let the fully inflated tire sit unmounted for 12 to 24 hours. This will allow any leaks to show up and will let the tire expand. Refill the tire with air if necessary and remount on the aircraft.
11. Complete the logbook entries. (See Chap. 8.)

7

Propeller Repair and Care

Aircraft Propellers

Wood Propellers

Most fixed pitch props are made of wood or aluminum. (See Fig. 7-1.) Wooden props are usually mounted on older airplanes or on the numerous experimental aircraft being constructed. Although wooden props are predominantly a fixed pitch design, some older, high-performance airplanes actually sported constant-speed wood propellers! Mahogany, cherry, oak, and birch are the most commonly used woods to construct propellers, although birch is by far the leading choice. Construction of a propeller starts with a layer of wood, or plank, which is then glued and bonded to another layer. A minimum of at least five layers of wood is used during construction. Once the planks are glued together, they become what is referred to as a *blank*. The blank is then rough-cut into a propeller shape and allowed to sit for approximately a week to 10 days to allow any moisture to permeate evenly throughout the wood. The blank is now upgraded to a *white*. The white is then precision-cut to the exact pitch required for the prop's future application—in other words, is the propeller destined to be a climb

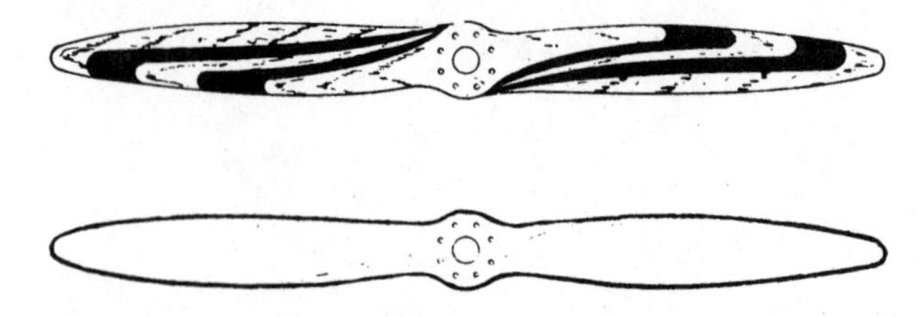

7-1 *Typical general aviation propellers.*

prop, or a cruise prop, or some combination of the two? Once this is accomplished, the hub is drilled for attachment bolts, and the propeller is coated in plastic to protect the finish. Not all props are plastic-coated, however. Some have a cotton fabric doped to the tips or a metal tip is added to the last 12 in of the leading edge of the blades for protection. Three $\frac{3}{16}$-in holes are drilled near the tip of each blade to allow moisture an escape route and allow the prop to "breathe." The prop is now ready for installation.

Aluminum Propellers

Aluminum has all but replaced wood propellers on most production aircraft. Aluminum props are thinner, lighter, and much more durable than wood. The manufacturing process isn't anywhere near as romantic, however; there are no blanks or whites, just aluminum alloys machined into the desired shape and twisted to the desired pitch. The finished prop is balanced and the surface coated with paint or an anodizing process.

Propeller Terms

The propeller takes the power being developed by the engine and turns it into motion or thrust. Propellers don't just move air backward, they actually produce lift which pulls or pushes the aircraft forward depending upon the aircraft's configuration. The following terms are helpful for understanding how the propeller operates (see Fig. 7-2):

Angle of attack. Just like the wings on an airplane, the propeller has an angle of attack. It's the angle between the chord line of the propeller blade and the relative wind.

Back. The curved section of the propeller blades as viewed from the front.

Blade. The section of propeller that starts at the prop hub and extends to the tip.

Blade angle. The angle between the blade's chord line and the prop's plane of rotation.

Blade root. The portion of the propeller blade closest to the prop hub.

Boss. The center part of a fixed-pitch propeller.

Constant speed prop. The system that utilizes a propeller governor to maintain a selected RPM by adjusting the blade angle.

Effective pitch. The total forward distance the aircraft travels for each revolution of the propeller.

Face. The flat side of the propeller blade. It's the area of the blades visible from the cockpit.

Fixed-pitch prop. A propeller that has a fixed blade angle and is not adjustable.

Hub. The center of a propeller assembly that

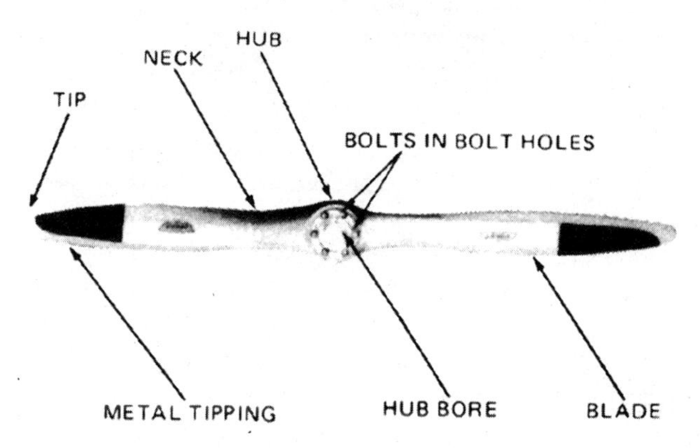

7-2 *Parts of a propeller.*

is mounted to the engine crankshaft; the propeller blades are attached to the hub.

Pitch. Also known as *blade angle.*

Shank. The thick middle section of a propeller blade.

Tip. The end of the propeller blade.

There are two basic types of propellers widely used on general aviation aircraft today. The fixed-pitch propeller and the controllable pitch propeller.

Fixed-Pitch Propellers

The fixed pitch prop is usually found on noncomplex general aviation airplanes. It can be made of laminated wood or aluminum and is normally constructed as one piece. As the name implies, the pitch of the blades cannot be changed.

McCauley and Senseninch propellers are the most commonly used on light aircraft. Each manufacturer has its own designation system, but the systems are similar enough that knowledge of one can easily be transferred to others. Take the McCauley designation lB90/CM7246—the *1B90* is the basic prop design number. *CM* refers to which engine crankshafts the prop will fit. The *72* indicates that the prop is 72 in in diameter, and the *46* is the pitch of the blades in inches at the 75 percent blade station. Now look at Senseninch's designation nomenclature. A Senseninch 76DM6S5-2-54 has a diameter of 76 in; *DM6* is indicative of blade design and which hub the prop will fit on. The *2* indicates that the diameter is actually 2 in less than the design, so actually we have a 74-in diameter prop with a blade pitch of 54 in, again at the 75 percent station.

Constant-Speed Propellers

Within a certain engine operating range, a constant-speed prop will remain at a preset RPM. (See Fig.

7-3.) This is accomplished by varying the propeller's blade angle. The blade angles are changed by the propeller governor.

Constant-speed propellers operate on a fixed-force principle. This force is created by centrifugal force acting against a spring or counterweights any time the propeller is turning. In order to change the pitch of the prop blades, this centrifugal force has to be overcome. To do that, a variable force must be applied, and this is accomplished by utilizing engine oil pressure. The prop governor maintains a selected RPM by varying oil pressure. The oil pressure controls the prop governor through the propeller control lever. In order to use oil pressure to move the propeller blades, the pressure has to be boosted or increased in some way. This is the function of the oil booster pump located in the prop governor. (See Fig. 7-4.) The prop governor has a drive shaft which is connected to the engine

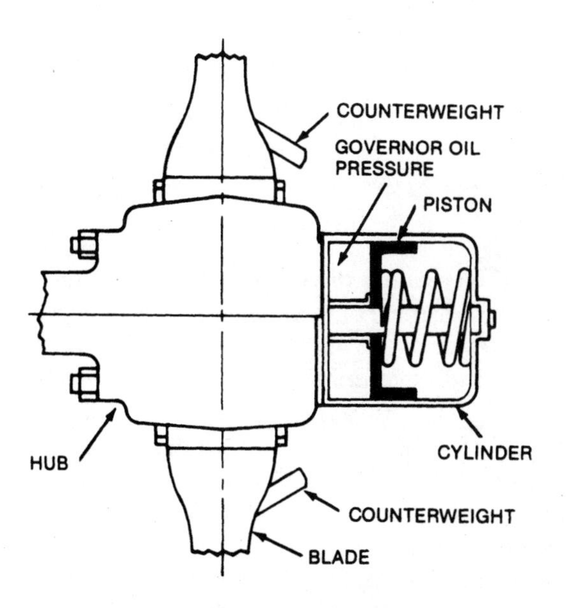

7-3 *Typical constant-speed propeller operation. (Courtesy of Hartzell Propeller)*

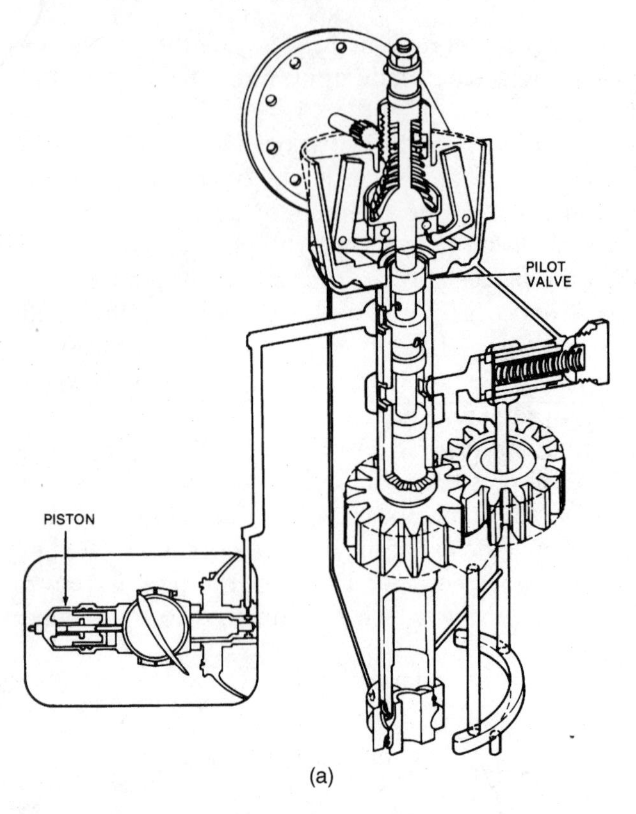

7-4 *Position of governor flyweights in response to engine*

crankshaft. The governor drive shaft rotates at a preset percentage of the speed of the engine crankshaft, usually around 80 to 110 percent. The oil booster pump is connected to the governor drive shaft and takes the engine oil and increases the pressure to a valve that the propeller can use. This boosted oil is then directed into an oil passage that is routed through the center of the prop governor. On each side of this passageway are open ports which correspond to a certain prop setting. A pilot valve moves up and down this passageway, opening and closing ports, which in turn vary the blade angle.

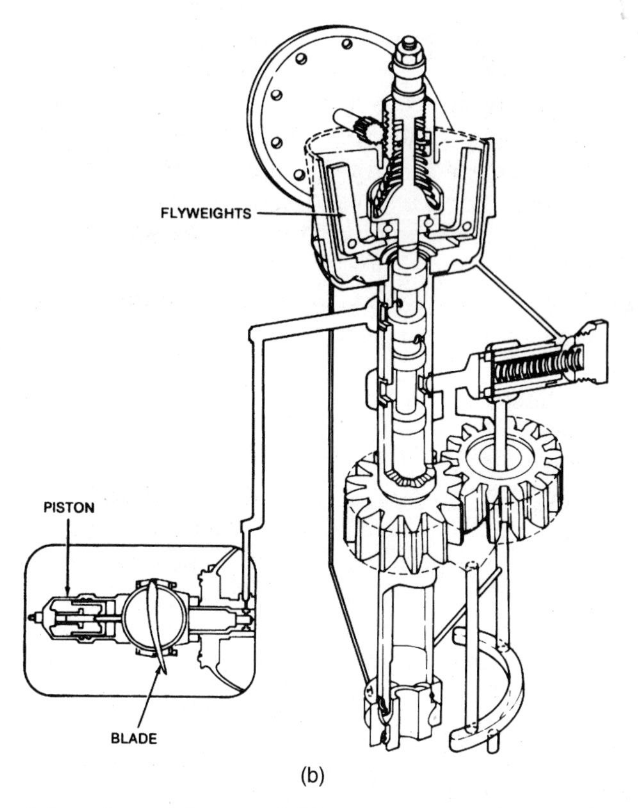

(a) underspeeding and (b) overspeeding. (Courtesy of Hartzell Propeller)

Inside the prop dome on the front of the propeller are flyweights that are attached to the drive shaft. The flyweights determine the position of the pilot valve. These flyweights tilt outward when prop RPM increases and tilt inward when prop RPM decreases. This inward and outward movement of the flyweights determines which way the pilot valve will move. As the flyweights tilt inward (RPM decreases), the pilot valve is lowered. As the flyweight tilts outward (RPM increases), the pilot valve is raised. The pilot valve is moving in relationship to the prop RPM, and this movement is what allows the oil to flow under pressure into the ports that adjust the blade angle. Controlling the movement of the flyweights is accomplished by a speeder spring which is located above the flyweights. The compression of the spring is controlled by utilizing the prop control lever inside the cockpit. By moving the prop control forward, more RPM's are called for, and this is accomplished by the speeder spring being compressed. The compressed spring applies pressure to the flyweights, which tilt inward in response. As the flyweights tilt inward, the pilot valve is lowered. This action decreases the blade angle and increases the prop RPMs. Soon the centrifugal force of the flyweights will overcome the speed springs compression, and the pilot valve will return to a neutral position. This, in effect, locks in the RPM setting. It's easy to deduce from the above that the opposite action will occur if the prop control lever is pulled aft. The speed spring compression will be reduced, decreasing the tension on the flyweights. They, in turn, will tilt outward causing the pilot valve to be raised, and the blade angle will increase and RPMs will decrease, until the centrifugal force on the flyweights decreases and the pilot valve returns to a neutral position. The flyweight and the pilot valve will react in the same manner regardless of the cause. In other words, the prop doesn't know if it's the pilot moving the prop control lever or if it's a steep climb creating enough drag to slow the propeller RPMs. The system will still do its job.

Look at a climb for example. The pilot pulls the nose of the airplane up and wants to exchange airspeed for altitude without increasing power. As the nose comes up, the airspeed decreases, and the angle of attack of the propeller blade will increase. This causes an increase in drag and slows down the propeller RPMs. The reduced centrifugal force on the flyweights allows them to tilt inward. This lowers the pilot valve, and the blade angle is reduced, resulting in an increase in RPMs. Once the RPMs reach their original setting, the pilot valve returns to neutral.

If the pilot were to begin a descent, the opposite reaction would occur. As the airspeed builds up, the prop would turn faster. The prop governor would recognize this increase in RPMs by the increased centrifugal force on the flyweights. They would react by tilting outward, and the pilot valve would be raised. This would increase the blade angle and decrease the RPMs. Once the RPMs returned to their original setting, the pilot valve would return to neutral. Changing the throttle will have the exact same effect on the prop RPMs. Increasing the throttle will cause the blade angle to increase, thus preventing the RPMs from exceeding their setting, and decreasing the throttle would cause a decrease in blade angle.

Propeller Maintenance Tips

Use the following tips and procedures to inspect propellers.

Wood or Composition Propellers and Blades

Owing to the nature of the wood itself, wood propellers and blades must be inspected frequently to assure continued airworthiness. Inspect for such defects as cracks, bruises, scars, warpage, evidence of glue failure and separated laminations, broken-off sections, and defects in the finish. Composition blades must be handled with the same consideration as wood blades.

Fixed-Pitch Propellers

Fixed-pitch propellers are normally removed from the engine at engine overhaul periods. Whenever the propeller is removed, visually inspect the rear surface for any indication of cracks. If any indications are found, disassemble the metal hub from the propeller. Inspect the bolts for wear and cracks at the head and threads and, if cracked or worn, replace with new (AN) bolts also. Inspect for elongated boltholes, enlarged hub bore, and check for cracks inside of bore or anywhere on the propeller. Repair propellers found with any of these defects. If no defects are found, the propeller may be reinstalled on the engine. Prior to installation, touch up with varnish any places where the finish has worn thin, become scratched or nicked.

On new fixed-pitch propeller installations, inspect bolts for tightness after the first flight and after the first 25 hours of flying. Always inspect and check the bolts for tightness at least every 50 hours. No definite time interval can be specified, since bolt tightness is affected by changes in the wood caused by the moisture content in the air where the airplane is flown and stored. During wet weather, some moisture is apt to enter the propeller wood through the drilled holes in the hub. The wood swells, and, since expansion is limited by the bolts extending between the two flanges, some of the wood fibers become crushed. Later, when the propeller dries out during dry weather or due to heat from the engine, a certain amount of propeller hub shrinkage takes place, and the wood no longer completely fills the space between the two hub flanges. Accordingly, the hub bolts become loose.

In-flight Tip Failures

In-flight tip failures may be avoided by frequent inspections of the metal cap and leading edge strip and the surrounding areas. Inspect for such defects as looseness or slipping, separation of soldered joints, loose screws, loose rivets, breaks, cracks, eroded sections, and corrosion. Inspect for separation between the metal leading edge and cap, which would indicate the cap is moving outward in the direction of centrifugal force. This condition is often accompanied by discoloration and loose rivets. Inspect the prop tip for cracks by grasping it by hand and slightly twisting about the longitudinal blade centerline and by slightly bending the tip backward and forward. If leading edge and cap have separated, carefully inspect for cracks at this point. Cracks usually start at the leading edge of the blade. A fine line appearing in the fabric or plastic may indicate a crack in the wood.

Examine the wood close to the metal sleeve of wood blades for cracks extending outward on the blade. These cracks sometimes occur at the threaded ends of the lag screws and may be an indication of internal cracking of the wood. Check the tightness of the lag screws, which attach the metal sleeve to the wood blade. Inspect and protect the shank areas of composition blades next to the metal sleeve in the same manner as for wood blades.

Metal Propellers and Blades

Metal propellers and blades are generally susceptible to fatigue failure resulting from concentration of stresses at the bottoms of sharp nicks, cuts, and scratches. It is very important to inspect them frequently and carefully for such injuries. Propeller manufacturers publish service bulletins and instructions which tell how these inspections should be accomplished.

Inspect controllable pitch propellers frequently to determine that all parts are lubricated properly. It is especially recommended that all lubrication be accomplished at the intervals specified by the propeller manufacturer.

Repair of Wood and Composition Propellers

Wood Propellers

Carefully examine any wood propellers and blades requiring repair to be sure that they can be restored to their original airworthy condition. Refer doubtful cases back to the manufacturer. Carefully evaluate any propeller damaged in the following manner prior to attempting a repair:

A crack or deep cut across the grain of the wood
Split blades

Separated laminations, except the outside laminations of fixed-pitch propellers

More screw or rivet holes, including holes filled with dowels, than used to attach the metal leading edge strip and tip

An appreciable warp

An appreciable portion of wood missing

A crack, cut, or damage to the metal shank or sleeve of blades

Broken lag screws which attach the metal sleeve to the blade

Oversize shaft hole in fixed-pitch propellers

Cracks between the shaft hole and boltholes

Cracked internal laminations

Excessively elongated boltholes

Fill any small cracks that run parallel to the grain with glue. Thoroughly work the glue into all portions of the cracks. Allow it to dry, then sand it smooth and flush with the surface of the propeller. This also works with small cuts. Any dents or scars which have a rough surface and will hold a filler without inducing failure may be filled with a mixture of glue and clean fine sawdust thoroughly worked and packed into the defect. After the mixture has dried, sand it smooth and flush with the surface of the propeller.It is very important that all loose or foreign matter be removed from the defect so that a good bond of the glue to the wood is obtained.

Blade Inlays

Always make inlays of the same type of wood as the propeller blade: Use a yellow birch inlay with a yellow birch propeller, and try to obtain inlays as near as the same specific gravity of the propeller blade as possible. Make all repair joints to conform with Fig. 7-5, for a taper of 10:1, starting from the deepest point to the feather edge or the end of inlay. All measurements should be taken along a straight line parallel to the grain or general slope of the surface on the thrust and camber face. This also applies when making edge repairs. Try to extend the grain of the inlays in the same direction as the grain of the propeller laminations. Use a fishmouth, scarf, or butt joint when making inlays. A fishmouth joint will last longest, followed by a scarf or butt joint. Avoid using dovetail-type joints. Do not exceed 1 large, 2 medium, or 4 small widely separated inlays per blade. Do not overlap a trailing and a leading edge inlay more than 25 percent. Propeller blades of a thickness ratio of 0.12 or more may be repaired, provided the cross-grained cut does not exceed 20 percent of the chord in length and the depth of cuts does not exceed one-eighth of the blade thickness at the deepest point of damage. Blades with airfoil sections less than 0.12 thickness ratio may be repaired if the maximum depth of damage does not exceed one-twentieth of the blade section thickness at the deepest point of damage. To determine the thickness ratio for a propeller blade, divide the maximum thickness of the airfoil section by the blade chord at the three-fourths radius station (see Examples 7-1 and 7-2).

Example 7-1

Given	6 in	Chord
	0.72 in	Maximum thickness of airfoil section
Computation	0.72 ÷ 6 = 0.12 thickness ratio	

Example 7-1 shows the blade thickness ratio to be 0.12; therefore, an inlay one-eighth of blade thickness at the point of damage may be inserted.

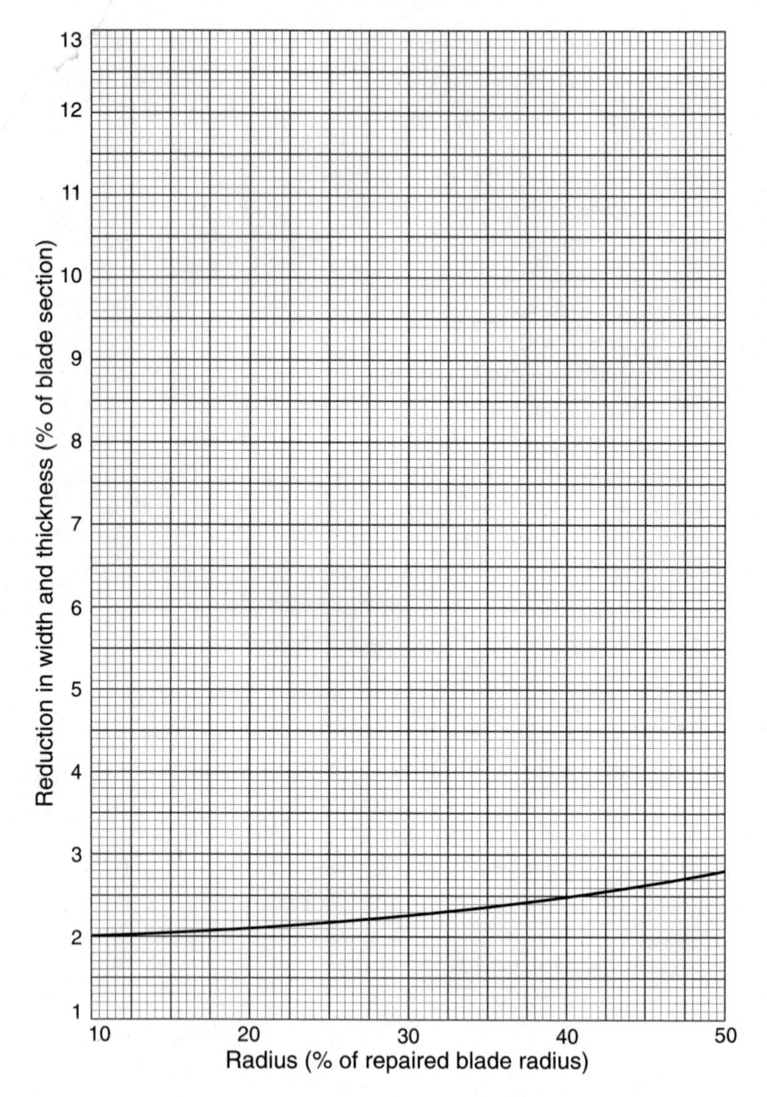

7-5 *Repair limits to blade section width and thickness for aluminum alloy propellers. Reductions shown are the maximum allowable below the minimum dimensions required by the blade drawing and blade manufacturing specification.*

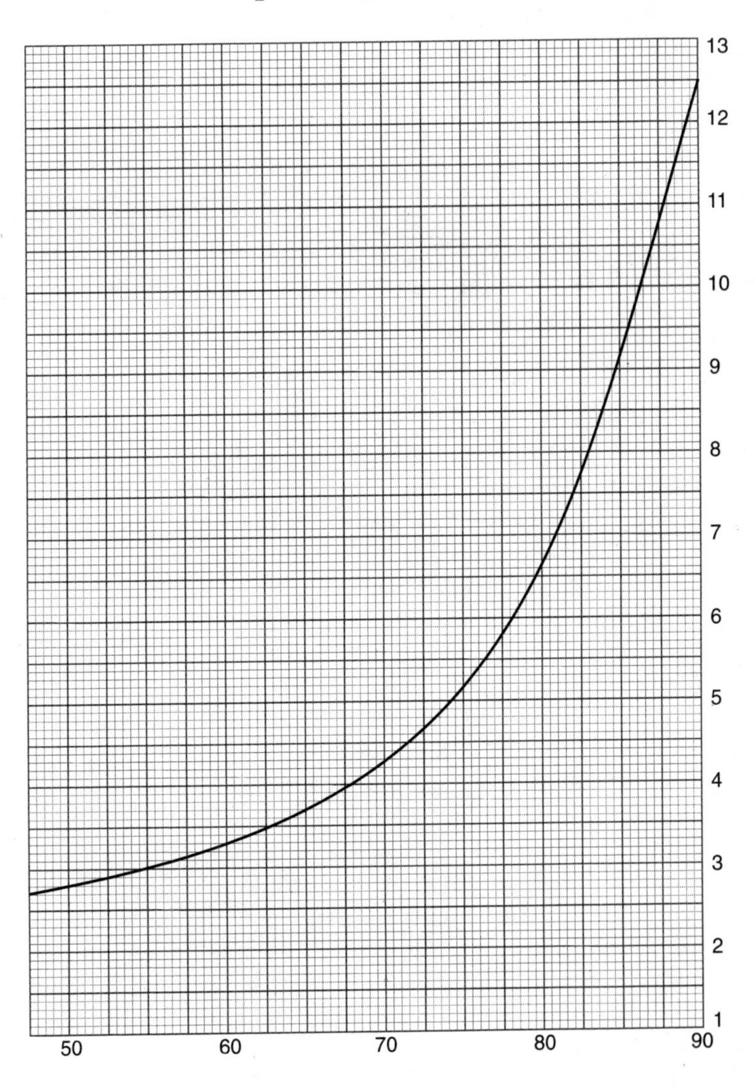
13
12
11
10
9
8
7
6
5
4
3
2
1
50
60
70
80
90

Example 7-2

Given	4 in	Chord
	0.36 in	Maximum thickness of airfoil section
Computation	0.36 ÷ 4 = 0.09 thickness ratio	

Example shows blade thickness ratio to be less than 0.12; therefore, an inlay one-twentieth of the blade thickness at the point of damage may be inserted.

Hub Inlays

Inlays in the sides of hubs of fixed-pitched propellers which would exceed in depth a value greater than 5 percent of the difference between the hub and bore diameters are not acceptable. In the portion of the blade where it fairs into the hub, allowable depths for inlays depend on the general proportions. Where the width and thickness are both very large in proportion to the hub and blade, maximum inlay depths of 7½ percent of the section thickness at the inlay are permissible. Where the width and thickness are excessively small, maximum inlay depths of 2½ percent of the section thickness at the center of the inlay are permissible; for propellers over 50 hp, cuts 2½ percent deep may be filled with glue and sawdust, whereas for propellers under 50 hp, cuts 5 percent deep may be filled with glue and sawdust.

Blade Profiling

Narrow slivers up to ⅛ in wide broken from the trailing edge at the wider portions of the blade may be repaired by sandpapering a new trailing edge, removing as little material as possible, and fairing in a new trailing edge using a smooth contour. Narrow both blades by the same amount. Near the hub or tip, use an inlay which does not exceed, at its greatest depth, 5 percent of the chord.

Propeller Tip Repairs

In order to replace the wood worn away at the end of the metal tipping, remove enough of the metal to make the minimum repair taper 10:1 each way from the deepest point. Because of the convex leading edge of the average propeller, this taper usually works out to approximately 8:1. Repairs under the metal tipping must not exceed 7½ percent of the chord for butt or scarf joints and 10 percent for fishmouth joints, with ¾-in maximum depth for any repair.

The scarfing of wood tips onto a propeller blade to replace a damaged tip is not considered an acceptable repair. The success of this type of repair fully depends on the strength and quality of the glue joint. Since it is difficult to apply pressure evenly over the glue area, and since no satisfactory means are available for testing the strength of such joints, it is quite possible for defective glue joints to exist and remain undetected until failure occurs.

Lamination Repairs

Whenever the glue joint of an outside lamination of fixed-pitch propellers is open, repair the propeller by removing the loose lamination and gluing on a new lamination of kiln-dried wood of the same kind as the original lamination. It is not usually economical to attempt to repair separations between other laminations. Repair outside laminations, which have been crushed at the hub by excessive drawing up of hub bolts, by planing and sanding one hub face smooth, removing a lamination on the other hub face, and replacing it with a new lamination, thus building the hub thickness up to the original thickness. It is permissible to replace both outer laminations if necessary and feasible.

Repair of Elongated Boltholes

It is permissible to repair elongated boltholes by the insertion of a steel bushing around each bolt.

Machine the inside diameter of the bushing to fit the bolt snugly, and the outside diameter should be approximately ¼ in larger than the bolt size. Make the bushing approximately ½ in long. Drill the face of the hub with a hole concentric with the bolthole and only to a sufficient depth to accommodate the bushing so that it does not protrude above the surface of the wood hub. Do not drive the bushing into the hub. Fit the bushing into the hole in the hub with a clearance not exceeding 0.005 in after moistureproofing. Protect the bushing hole from moisture by two coats of aluminum paint, varnish, glue, or other moisture-resistant coating.

Repair of Plastic-Covered Propellers and Blades
Repair small cracks, dents, scratches, and cuts in the plastic of plastic-covered wood propellers and blades by using special repair cement supplied by the manufacturer. Instructions accompany the cements. Polishes and cleaners are available for preserving the gloss finish of varnished or plastic-covered propellers and blades.

Repair of Composition Blade

Repair composition blades in accordance with the manufacturer's instructions. For repairs to the metal cap and leading edge strip, follow the methods and procedures discussed in the following paragraphs.

Fabric Replacement.
Replace fabric used to strengthen the tips of wood blades when it becomes loose or worn through. Launder the fabric (mercerized cotton airplane cloth) to remove all sizing. Cut a piece of fabric to approximate size required to cover both faces of outer portion of blade. Recover the same portion that the original fabric covered. Apply a thick solution of glue to the wood where the fabric is to be put. Put the fabric on the glued

surface, starting at the leading edge of the thrust face, and work toward the trailing edge. Fold the fabric around the trailing edge over the camber face and toward the leading edge. Make a joint on the leading edge where it will be covered by the metal tipping. As the fabric is put on, smooth it out over the wood to prevent air bubbles or uneven gluing. The fabric must be perfectly flat on the blade. Trim any excess fabric off with small scissors. *Do not* cut or score the fabric with a knife. Allow the glue under the fabric to dry about 6 hours, then brush two coats of nitrate dope on the fabric, allowing it to dry for ½ hour. Then sand the fabric lightly and brush a coat of pigment dope over it. Lightly sand the uncovered portion of the wood and apply two coats of a good grade of moisture resistant varnish, allowing 12 to 16 hours drying time between coats.

Metal Tip Replacement.
Always replace the tipping when it cannot be properly repaired. Small cracks in the narrow necks of metal between pairs of lobes of the tipping are to be expected and are not considered defects. All other cracks should be considered defects that require repair or elimination by new tipping. If the propeller does not require fabric, apply two coats of varnish to the wood to be covered by the metal tipping. If new fabric has been applied, puncture it with a pointed tool at each screw and rivet hole. Apply varnish, white lead, aluminum paint, etc., to all holes, allowing the wood to absorb as much as it can. Be sure to use the original screw and rivet holes in the propeller. New holes are not to be drilled. A number of wood propeller tip failures have occurred which have been attributed to the practice of drilling new rivet and/or screw holes in the wood tips when replacing the metal tipping. To avoid this problem, we strongly recommend that the manufacturer's procedure be closely adhered to, and any procedure which

involves drilling new holes in the wood tip and plugging old holes with dowels is to be discontinued immediately.

Once new tips and leading-edge strips are obtained, cut them to size and form them to the approximate shape of the leading edge of the propeller. These pieces are usually supplied without holes so they can be drilled when lined up with the old screw and rivet holes in the propeller. If new material is not available, the old tipping can be hammered flat and used as a pattern to lay off a new tip. For this purpose, use a piece of sheet metal of the same material and thickness as the old tip, and remove the burr from the cut edges of this piece.

Next, lay the cutout flat metal strip over the leading edge, and bend this metal down over the leading edge of the propeller, being careful that the metal extends an equal width on thrust and camber faces. This can be done by following the impressions of the old tipping lines. Numerous waves will occur in the metal, but these will be slowly eliminated as the work progresses. Obtain several pieces of strong rubber tape, 4 ft long, ½ in wide, and 1⁄16 in thick. While forming the metal, hold it in place on the propeller by wrapping the rubber tape around the blade. Start at the tip and work inboard, being careful not to cover any pencil lines placed on the propeller which show the location of the rivet holes. While metal is held in place, tap the leading edge with a rubber mallet, using moderate force to make sure the metal is seated against the wood along the nose of the leading edge. Smooth the metal by hammering it with the mallet. Back up the opposite side of the blade with a laminated hardwood bucking block with an iron weight built in the center. The block should measure about 2 by 4 in. Start at the end of the blade and work toward the hub, moving the bucking block so that it is always under the section being hammered. Continue to do this until the metal is well shaped to the profile of the propeller. Check to see that the metal has not moved from its original position.

With a center punch and a hammer, locate the old screw and rivet holes. Punch the metal approximately 1/4 in from the edge. After all holes have been located, remove the metal from the propeller. Drill the screw and rivet holes in the metal with a 1/8-in drill. File off burrs on the inside of the metal. Run the drill through the original rivet holes in the propeller in order to clean them. Cut or saw slots in the metal at the original positions. (Refer to old tipping metal for location of the slots.) Place the metal leading edges on the blades they were formed to fit and hold them in place with rubber tape. With a center punch as large as or slightly larger than the diameter of the screw and rivet heads, punch metal into the original countersunk holes in the wood so that the screw and rivet heads may be entered to the correct depth (not more than 1/32 in below the surface of the metal). Use screws and rivets of the appropriate material. Use screws one size larger than were originally in the propeller and rivets of the solid flat countersunk-head type. Insert screws and rivets in their respective holes, and install rivets with their heads on the thrust face of the propeller. After rivets are tapped in place, cut off the excess length of the rivet, leaving 1/8 in for heading. When an assistant backs up the rivets with a steel bar 18 in long and pointed to fit the rivet head, hammer the rivets either by hand or with a pneumatic hammer. Screws may be driven either by hand or with an electric screwdriver.

Cut the metal of the cap tip on the camber face of the propeller to the shape of the propeller tip. Bevel the edge by hand filing and trim off flat side of metal cap so that it extends about 1/16 in all around the tip of the propeller. Form a hardwood block to the shape of the propeller tip thrust face. Put metal tipping in place and clamp this block to the underside of the tip with a C clamp; turn this 3/16 in of metal up and over the camber face of the tip. Tighten and complete the lap joint. Mount the propeller blade solidly, with the thrust face up, on a stand supporting the blade at several points along its radius. With a hammer and a flat-faced tool, proceed to smooth the metal, starting at the nose of the leading edge and working toward the edge of the metal until all wrinkles and high spots are removed. At the edge, use a caulking tool and, in the same manner, press the metal edge tightly against the wood. Turn the propeller over and repeat this operation on the camber face. Make sure that the thin tip is supported at all times when hammering. Apply solder over rivet and screw heads and over the metal seam of the tip of the propeller. Use 50-50 solder in wire form. Use muriatic acid as flux when soldering brass. Use stainless steel soldering flux when soldering stainless steel tipping. File excess solder off and check the propeller balance while doing so. Polish the metal with a fine emery cloth or an abrasive drum driven by a flexible shaft. Vent the tipping by drilling three holes, using a no. 60 drill (0.040), 3/16-in deep in the tip end. Drill ventholes parallel to the longitudinal axis of the blade.

Refinishing

After repairing a blade, it is usually necessary to refinish it. In some cases it may be necessary to completely remove the old finish. Apply any finish according to the recommendations of the propeller manufacturer. Refinishing of plastic-covered blades requires special techniques. Some manufacturers make this type of information available through service bulletins. Refinish or repaint the wood blades carefully so that the balance of the entire propeller is not disturbed. Coating one blade heavier than the other will produce an unbalanced blade and cause a noticeable vibration during flight.

Propeller Balancing

It is always necessary to check the balance of the propeller after any repairs or refinishing. Accomplish the final balance on a rigid knife-edge balancing stand or on a suspension-type balancer in a room free from any air currents (see Fig. 7-6). The blades should not show any persistent tendency to rotate from any position on the balance stand, or to tilt on the suspension balancer. Any horizontal unbalance tendencies may be corrected by the application of a finish or solder to the light blade. The light blade may be coated with a high grade of primer allowing for a finishing coat. After allowing each coat to dry 48 hours, recheck the balance. Then, either remove as much finish as required by carefully sandpapering or add more by applying an additional finishing coat. Recheck the balance and sandpaper or apply additional finish as may be required to effect final balancing. To correct vertical imbalance in fixed-pitch propellers, apply putty to the light side of the wood hub at a point on the circumference approximately 90 degrees from the longitudinal centerline of the

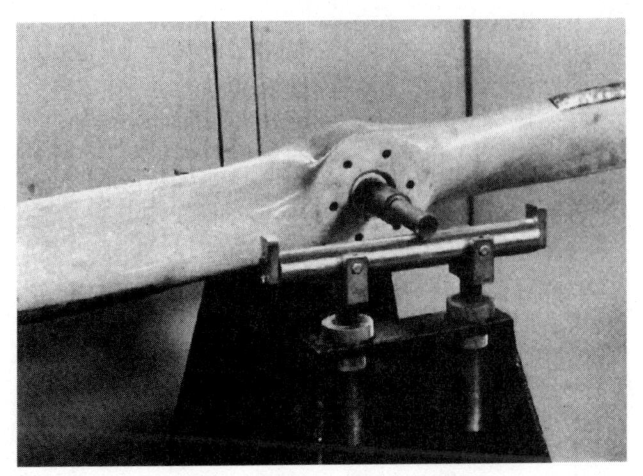

7-6 *Propeller on a balance stand.*

blades. Weigh the putty and prepare a brass plate weighing slightly more than the putty. The thickness of the plate should be from 1/16 in to 1/8 in, depending on the final area, and be of sufficient size to accommodate the required number of flathead attaching screws. The plate may be made to fit on the hub face or to fit the shape of the light side of the wood hub and drilled and countersunk for the required number of screws. Attach the plate and tighten all of the screws. After the plate is finally attached to the propeller, secure the screws to the plate by soldering the screw heads and recheck the propeller balance. All edges of the plate may be beveled to reduce its weight as necessary. Do *not* drill holes in the propeller or insert lead or other material to assist in balancing.

Repair of Metal Propellers

Because of the critical effects of surface injuries and their repair on the fatigue life of steel blades, all repairs must be made in accordance with the manufacturer's instructions. Aluminum-alloy propellers and blades with dents, cuts, scars, scratches, nicks, leading edge pitting, etc., may be repaired, provided the removal or treatment does not materially affect the strength, weight, or performance of the blade. More than one injury is not sufficient cause alone for rejection of a blade. A reasonable number of repairs per blade may be made and not necessarily result in a dangerous condition, unless their location with respect to each other is such to form a continuous line of repairs that would materially weaken the blade. Suitable sandpaper or fine-cut files may be used for removing the necessary amount of metal. In each case, the area involved will be smoothly finished with no. 00 sandpaper, and each blade from which any appreciable amount of metal has been removed will be properly balanced before it is used. To avoid removal of an excessive amount of metal, local etching should be accomplished at intervals during the process of removing suspected cracks. Upon completion of the repair, carefully inspect the entire blade by etching or anodizing. Remove all effects of the etching process with fine emery paper. Blades identified by the manufacturer as being cold-worked (shot-blasted or cold-rolled) may require peening after repair. Accomplish all repair and peening operations on this type of blade in accordance with the manufacturer's instructions. However, do not peen down the rough edges of any damage in order to lap metal over the injury.

Round out nicks, scars, cuts, etc., occurring on the leading edge of aluminum alloy blades. Blades that have the leading edges pitted from normal wear in service may be reworked by removing sufficient material to eliminate the pitting. In this case, remove the metal by starting well back from the edge and working forward over the edge in such a way that the contour will remain substantially the same, avoiding any abrupt changes in contour. The trailing edges of the blades may be treated in substantially the same manner. On the thrust and camber face of the blades, remove the metal around any damage such as dents, cuts, scars, scratches, nicks, or longitudinal surface cracks. Removing the metal will form shallow saucer-shaped depressions. Be sure to remove the deepest area of damage and any raised metal around the edges of the damage. For repaired blades, the permissible reduction in width and thickness for minimum original dimensions allowed by the blade drawing and blade manufacturing specifications are shown in Fig. 7-5. These are for locations on the blade from the shank to 90 percent of the blade radius. Beyond the 90 percent blade radius point, the blade width and thickness may be modified as per manufacturer's instructions.

Shortening Blades

If the removal or treatment of defects on the tip necessitates shortening a blade, shorten each blade the same amount. Mark the shortened blades to correspond with the manufacturer's system of model designation to indicate propeller diameter. When making the repair, do not reduce the propeller diameter below the minimum diameter limit shown on the pertinent airplane specification.

Straightening Propeller Blades

Always repair bent blades in accordance with the manufacturer's instructions. First, carefully check the extent of a bend in face alignment by means of a protractor. Only bends not exceeding 201 at 0.15-in blade thickness to 0 degrees at 1.1-in blade thickness may be cold-straightened. Blades with bends exceeding these values should be repaired only at facilities using proper heat treatment equipment. Always inspect the blades for cracks and other injuries both before and after straightening.

Repair Tolerances

The following tolerances are those listed in the blade manufacturing specification for aluminum alloy blades and govern the width and thickness of new blades. These tolerances are to be used with the pertinent blade drawing to determine the minimum original blade dimensions to which the reduction may be applied. When repairs reduce the width or thickness of the blade below these limits, the blade must be rejected. The face alignment or track of the propeller should fall within the limits recommended by the manufacturer for new propellers (see Table 7-1).

No repairs are permitted to the shanks (roots or hub ends) of aluminum alloy adjustable-pitch blades. The shanks must be within manufacturer's tolerances.

Table 7-1. Propeller Blade Dimensions and Tolerances

Basic diameter less than 10′6″	
Blade width	
From shank to 24″ station	±3/64
From 24″ station to tip	±1/32
Blade thickness	±0.025
Basic diameter 10′6″ to less than 14′0″:	
Blade width	
From shank to 24″ station	±1/16
From 30″ station to tip	±1/32
Blade thickness	
From shank to 24″ station	±0.030
From 30″ station to tip	±0.025
Basic diameter 14′0″ and over	
Blade width	
From shank to 30″ station	±3/32
From 36″ station to tip	±1/16
Blade thickness	
From shank to 30″ station	±0.040
From 36″ station to tip	±0.035

Propeller Balance

Upon completion of repairs, check horizontal and vertical balance and correct any unbalance as recommended by the manufacturer. A coaxial hole is drilled in the butt end of certain aluminum alloy detachable blades for the insertion of lead to obtain static horizontal balance. The size of this hole should not be increased. For vertical balance, follow the manufacturer's specific instructions.

Propeller Hub and Flange Repair

Repairs to steel hubs and parts must be accomplished only in accordance with the manufacturer's recommendations. Welding and remachining is permissible only when covered by manufacturer's service bulletins.

When the propeller boltholes in a hub or crankshaft flange on fixed-pitch propellers become damaged or oversized, it is permissible to make repairs in accordance with the recommendations of the propeller metal hub manufacturer or the engine manufacturer as applicable. Obtain from the engine or propeller hub manufacturer suitable flange bushings with threaded or smooth bores. Drill the flange and insert the bushings as recommended by the engine manufacturer. Next, drill the rear face of the propeller to accommodate the bushings, and protect the holes with two coats of aluminum paint or other highly moisture-resistant coating. Use bolts of the same size as those originally used. Any of the following combinations may be used: (1) safety bolt and castellated nut, (2) safety bolt (drilled head) and threaded bushing, or (3) an undrilled bolt and self-locking nut. Where it is desirable to use oversized bolts, obtain suitable aircraft standard bolts 1/16-in larger than the original bolts. Enlarge the crankshaft propeller flange holes and the propeller hub holes sufficiently to accommodate the new bolts without more than 0.005-in clearance. Remember, such reboring will be permitted only once.

Control Systems

Components used to control the operation of certificated propellers should be inspected, repaired, assembled, and/or tested in accordance with the manufacturer's recommendations. Only those repairs which are covered by the manufacturer's recommendations should be made, and only those replacement parts which are approved under FAR 21 should be used.

Overhaul of Propellers

Hub and Hub Parts

First, disassemble the propellers slated for overhaul and clean all the hub parts in accordance with the manufacturer's recommendations. Next, make an inspection of the parts to determine that all the critical dimensions are within the manufacturer's specified tolerances. Take particular care to check the 90-degree relationship between the shaft bore and the blade socket centerline and the track of the blade sockets, as these are the dimensions which are most likely to be affected by accidents. Reject and replace any hub which is sprung, worn, or damaged. Stress risers such as cuts, nicks, or tool marks must be carefully stoned or the part rejected. Carefully inspect splines and cone seats for signs of wear. Cones and cone seats may show discoloration, pitting, and corrosion. Generally, corrosion and discolored spots may be removed by light lapping. Pitting is not grounds for rejection, as long as 75 percent of the bearing area is not affected and the pitted areas are well dispersed about the cone-bearing area. After cleaning, minutely inspect steel hubs and parts for cracks by the wet or dry magnetic particle method at every major overhaul period. It is not necessary to remove the plating or special external finish for this inspection unless required by the manufacturer's procedures.

Steel hubs which adapt fixed-pitch propellers to all taper crankshafts are susceptible to cracks along the keyway which often extend into the flange lightening holes. Carefully inspect these hubs by the magnetic particle method at the engine's overhaul period. Any crack is cause for rejection.

Inspection of Aluminum Propellers and Blades

Carefully inspect aluminum propellers and blades for cracks and other injuries. A transverse (chordwise) crack or flaw of any size is cause for rejection. Refer any unusual condition or appearance revealed by these inspections to the manufacturer. Some acceptable inspection methods are the acid-etching process, the anodizing process, the fluorescent penetrant process, and the dye penetrant process. Etching is accomplished by immersing the blade in a warm 20% caustic soda solution and cleaning the blade with a warm 20% nitric acid solution. Rinse the blade with warm water between the caustic bath and the acid bath. Also use a warm water rinse following the acid bath. Be sure to remove all the effects of the etching by polishing. Try to maintain the caustic and acid solution at a temperature of 160 to 180°F. The blade surfaces are then examined with a magnifying glass of at least three-power. A crack will appear as a distinct black line. The fluorescent penetrant method is recommended as a supplement to the caustic etch for the inspection of the shanks (roots or hub ends) of adjustable pitch blades.

Suspected cracks or defects should be repeatedly local-etched until their nature is determined. With a no. 00 sandpaper, clean and smooth the area containing the apparent crack. Apply a small quantity of caustic solution to the suspected area with a swab or brush. After the area is well darkened, thoroughly wipe it with a clean (dampened) cloth. Too much water may entirely remove the solution from a crack and spoil the test. If a crack extending into the metal exists, it will appear as a dark line or mark, and using a magnifying glass, small bubbles may be seen forming in the line or mark. Immediately upon completion of the final checks, remove all traces of the caustic solution by use of the nitric acid solution. Wash the blade thoroughly with clean (fresh) water.

The chronic acid anodizing process is superior to caustic etching for the detection of cracks and flaws. This process should be used for general inspection of the blades, for any material defects, or for final checking of repairs performed during overhaul. First, immerse the blades in the anodizing bath as far as possible. Next, follow the anodizing treatment by a rinse in clear, cold, running water from 3 to 5 minutes, and dry the blades as quickly as possible, using compressed air. Allow the dried blades to stand for at least 15 minutes before examination. Flaws (cold shuts or inclusions) will appear as fine black lines. Cracks will appear as brown stains caused by chromic acid bleeding out onto the surface. The blades may be sealed for improved corrosion resistance by immersing them in hot water (180 to 212°F) for one-half hour. Do *not* immerse the blades in hot water before the examination for cracks, because the heat will expand the cracks and allow the chromic acid to be washed away.

Assembly

Accomplish the assembly of the propeller hub and blade in accordance with the manufacturer's recommendations. Replace clevis pins, bolts, and nuts which show wear or distortion. Never use cotter pins and safety wire a second time. The use of self-locking nuts is permissible only where originally used or approved by the manufacturer.

Installing the Propeller to the Engine

Fixed-Pitch Propellers

Loose hub bolts cause elongated boltholes and damage to the hub bolts. If not corrected, the bolts can break off or friction can cause enough heat to be generated that it affects the glue and even chars the wood. After successive running, cracks start at the boltholes. These cracks are caused, or at least accentuated, by shrinkage of the wood caused by the excessive heat generated.

If allowed to progress, the propeller usually flies apart or catches on fire.

On some engines equipped with a crankshaft with an integral propeller hub flange, the outer edge of the lightening holes are at the same radius as the corresponding edge of the propeller hub boltholes. When inserting the bolts through the propeller, care must be exercised so that the bolts are inserted through the proper holes in the flange. Cases have been reported where the bolts were inserted through the larger lightening holes, and, accordingly, the bolt nuts bore only on the outer edges of the lightening holes. In such cases, continuous running of the propeller may cause the boltheads or nuts to slip off the flange and through the large openings in the flange, resulting in the subsequent loss of the propeller.

Both the conditions discussed above are very easy to detect and must be corrected immediately. In case the hub flange is integral with the crankshaft, first ascertain the bolts are properly installed, then make the inspection for bolt tightness in the same manner as for any other propeller hub. Use an open-end wrench to determine hub bolt tightness; if the nuts can be turned, remove the cotter keys and retorque the nuts to the desired setting. Tighten hub bolts, preferably with a torque wrench, to the recommended values which usually range from 15 to 24 ft-lb. If no torque wrench is available, an ordinary socket wrench may be used. This socket wrench should have a 1-ft extension lever and the wrench pulled up with the recommended force 12 in away from the center of the bolt which is being tightened. The tightening is best accomplished by tightening each bolt a little at a time, being sure to tighten alternate bolts which are diametrically opposite. Exercise care not to overtighten the hub bolts, thereby damaging the wood underneath the hub flanges. Avoid the practice of overtightening bolts to draw a propeller into track. Safety the nuts by means of cotter keys of proper size or heavy safety wire twisted between each nut. A continuous length of single safety wire is not acceptable, as wire failure will result in all nuts becoming unsafe.

Precautions

Blade Tip Identification

Numerous people have been fatally injured walking into whirling propellers. Painting a warning strip on the propeller serves to reduce chances of such injuries. Cover approximately 4 inches of the propeller tips on both sides with an orange yellow nonreflecting paint or lacquer. Be sure to open the drain holes in the metal tipping of wood blades after the tips have been painted.

Ground-Handling Practices

Wood propellers are especially susceptible to damage from improper handling. When moving an airplane, avoid bumping the propeller. The practice of pushing or pulling on a propeller blade to move an airplane should be avoided; imposing forces on a blade in excess of those for which the blade is designed is extremely easy. It is continually necessary to ascertain that the glue joints are in good condition and that the finish on the entire propeller will protect the propeller from absorbing moisture. Place two-bladed wood propellers, whether on or off an airplane, in a horizontal position to prevent unbalance from moisture absorption. A good precaution is to cover the propeller with a well-fitted waterproof cover when not in use. It is very important to protect the shank section of wood blades from moisture changes to prevent swelling and subsequent loosening in the metal sleeve. In the case of varnished blades, occasionally apply varnish around the shank at the junction of wood and metal. In the case of the plastic-covered blade, repair cement may be applied around the same joint.

8

ADs, FARs, and Maintenance Logbooks

Few pilots are familiar with the federal regulations regarding aircraft maintenance. Pilots and owners are as responsible for adhering to these rules as are the mechanics that work on the aircraft.

Recording Maintenance Approved for Pilots and Owners

This is the section that is of interest to most aircraft owners. It covers exactly what pilots or owners can maintain or replace on their aircraft without holding an airframe and powerplant license. Of course, there is a complete list of maintenance that a nonmechanic is allowed to perform. These items can legally be accomplished by the owner or pilot, but the person performing the work must have access to the aircraft service manual or FAA advisory circular 43.13-1. As long as the aircraft is not operated under FAR 121, 127, 129, or 135, FAR 43, Appendix A, provides a list of permissible maintenance.

Keep in mind, however, that all work must be entered in the appropriate aircraft logbook after it has been completed. Nothing will interest an FAA inspector more than, during a routine ramp check, discovering that the aircraft has new tires, seatbelts, etc., but the logbook doesn't reflect any maintenance. Failure to document maintenance is an excellent way to get an aircraft grounded. Even more serious are logbook entries for work which hasn't been accomplished. Do not accept an aircraft as airworthy if any item in the inspection process has been "pencil whipped." It's a rare occurrence, but sometimes in the rush to get an aircraft back in the air, especially one that works for a living, a mechanic may be tempted to sign off work that wasn't accomplished, explaining that the work was unnecessary or that he or she will get to it soon. Remember, anyone guilty of such an infraction can lose her or his certificate.

Preventive maintenance must be recorded in accordance with FAR 43.9 (later in this chapter). This is accomplished by recording in the appropriate aircraft maintenance logbook a description of the work performed. This should include entries indicating what work was accomplished and how it was done. A simple job like changing the oil should have a simple entry, for example: Changed oil with 5 quarts Aeroshell 15W-50. The entry must also have the date the work was performed and the signature of the person performing the task. This signature along with the person's certificate type and number constitutes approval for return to service.

Approved Tasks for Preventive Maintenance

Preventive maintenance is limited to the following work, provided it does not involve complex assembly operations:

1. Removal, installation, and repair of landing gear tires
2. Replacing elastic shock absorber cords on landing gear
3. Servicing landing gear shock struts by adding oil, air, or both
4. Servicing landing gear wheel bearings, such as cleaning and greasing
5. Replacing defective safety wing or cotter keys
6. Lubrication not requiring disassembly other than removal of nonstructural items such as cover plates, cowlings, and fairings
7. Making simple fabric patches not requiring rib stitching or the removal of structural parts or control surfaces; in the case of balloons, the making of small fabric repairs to envelopes (as defined in, and in accordance with, the balloon manufacturer's instructions) not requiring load tape repair or replacement
8. Replenishing hydraulic fluid in the hydraulic reservoir
9. Refinishing decorative coating of fuselage, balloon baskets, wings, tail group surfaces (excluding balanced control surfaces), fairings, cowlings, landing gear, cabin, or cockpit interior when removal or disassembly of any primary structure or operating system is not required
10. Applying preservative or protective material to components where no disassembly of any primary structure or operating system is involved and where such coating is not prohibited or is not contrary to good practices
11. Repairing upholstery and decorative furnishings of the cabin, cockpit, or balloon interior when the repairing does not require disassembly of any primary structure or operating system or interfere with an operating system or affect primary structure of the aircraft

12. Making small simple repairs to fairings, nonstructural cover plates, cowlings, and small patches and reinforcements (not changing the contour so as to interfere with proper airflow)
13. Replacing side windows where that work does not interfere with the structure or any operating system such as controls, electrical equipment, etc.
14. Replacing safety belts
15. Replacing seats or seat parts with replacement parts approved for the aircraft, not involving disassembly of any primary structure or operating system
16. Troubleshooting and repairing broken circuits in landing light wiring circuits
17. Replacing bulbs, reflectors, and lenses of position and landing lights
18. Replacing wheels and skis where no weight and balance computation is involved
19. Replacing any cowling not requiring removal of the propeller or disconnection of flight controls
20. Replacing or cleaning spark plugs and setting of spark plug gap clearance
21. Replacing any hose connection except hydraulic connections
22. Replacing prefabricated fuel lines
23. Cleaning or replacing fuel and oil strainers or filter elements
24. Replacing and servicing batteries
25. Cleaning of balloon bumer pilot and main nozzles in accordance with the balloon manufacturer's instructions
26. Replacement or adjustment of nonstructural standard fasteners incidental to operations
27. The interchange of balloon baskets and burners on envelopes when the basket or

burner is designated as interchangeable in the balloon-type certificate data and the baskets and burners are specifically designed for quick removal and installation

28. The installation of antimisfueling devices to reduce the diameter of fuel tank filler openings provided the specific device has been made a part of the aircraft "certificate data" by the aircraft manufacturer, the aircraft manufacturer has provided FAA-approved instructions for installation of the specific device, and installation does not involve the disassembly of the existing tank filler opening

29. Removing, checking, and replacing magnetic chip detectors

30. The inspection and maintenance tasks prescribed and specification identified as preventive maintenance in a primary category aircraft certificate or supplemental certificate holder's approved special inspection and preventive maintenance program when accomplished on a primary category aircraft provided:

- They are performed by the holder of at least a private pilot certificate issued under FAR 61 who is the registered owner (including co-owners) of the affected aircraft and who holds a certificate of competency for the affected aircraft (a) issued by a school approved under FAR 147.21 (f) of this chapter; (b) issued by the holder of the production certificate for that primary category aircraft that has a special training program approved under FAR 21.24 of this subchapter.
- The inspections and maintenance tasks are performed in accordance with instructions contained in the special inspection and preventive maintenance program approved as

part of the aircraft's type design or supplemental type design.

FAR Regulations

The following is a list of FAR regulations, from FAR 43, with which every pilot, owner, or homebuilder should be familiar (FAR 43.1 pertains to who is required to adhere to these regulations. Notice that experimental aircraft are exempt).

43.1 Applicability

(a) Except as provided in paragraph (b), this Part prescribes rules governing the maintenance, preventive maintenance, rebuilding, and alteration of any-
  (1) Aircraft having a U.S. airworthiness certificate;
  (2) Foreign-registered civil aircraft used in common carriage or carriage of mail under the provisions of Part 121, 127, or 135 of this chapter; and
  (3) Airframe, aircraft engines, propellers, appliances, and component parts of such aircraft.

(b) This Part does not apply to any aircraft for which an experimental airworthiness certificate as been issued, unless a different kind of airworthiness certificate had previously been issued for that aircraft.

43.2 Records of overhaul and rebuilding

(a) No person may describe in any required maintenance entry or form an aircraft, airframe, aircraft engine, propeller, appliance, or component part as being overhauled unless-

(1) Using methods, techniques, and practices acceptable to the Administrator, it has been disassembled, cleaned, inspected, repaired as necessary, and reassembled, and

(2) It has been tested in accordance with approved standards and technical data, or in accordance with current standards and technical data acceptable to the Administrator, which have been developed and documented by the holder of the type certificate, supplemental type certificate, or a material, part, process, or appliance approval under 21.305 of this chapter.

(b) No person may describe in any required maintenance entry or form an aircraft, airframe, aircraft engine, propeller, appliance, or component part as being rebuilt unless it has been disassembled, cleaned, inspected, repaired as necessary, reassembled, and tested to the same tolerances and limits as a new item, using either new parts or used parts that either conform to new part tolerances and limits or to approved oversized or undersized dimensions.

43.3 Persons authorized to perform maintenance, preventive maintenance, rebuilding, and alterations

(a) Except as provided in this section and §43.17, no person may maintain, rebuild, alter, or perform preventive maintenance on an aircraft, airframe, aircraft engine, propeller, appliance, or component put to which this Part applies. Those items, the performance of which is a major alteration, a major repair, or preventive maintenance, are listed in Appendix A.

(b) The holder of a mechanic certificate may perform maintenance, preventive maintenance, and alterations as provided in Part 65.

(c) The holder of a repairman certificate may perform maintenance and preventive maintenance as provided in Part 65.

(d) A person working under the supervision of a holder of a mechanic or repairman certificate may perform the maintenance, preventive maintenance, and alterations that his supervisor is authorized to perform, if the supervisor personally observes the work being done to the extent necessary to ensure that it is being done properly and if the supervisor is readily available, in person, for consultation. However, this paragraph does not authorize the performance of any inspection required by Part 91 or Part 125 of this chapter or any inspection performed after a major repair or alteration.

(e) The holder of a repair station certificate may perform maintenance, preventive maintenance, and alterations as provided in Part 145.

(f) The holder of an air carrier operating certificate or an operating certificate issued under Part 121, 127, or 135, may perform maintenance, preventive maintenance, and alterations as provided in Part 121, 127, or 135.

(g) The holder of a pilot certificate issued under Part 61 may perform preventive maintenance on any aircraft owned or operated by that pilot which is not used under Part 121, 127, 129, or 135.

(h) Notwithstanding the provisions of paragraph (g) of this section, the Administrator may approve a certificate holder under Part 135 of this chapter, operating rotorcraft in a remote area, to allow a pilot to perform specific preventive maintenance items provided

(1) The items of preventive maintenance are a result of a known or suspected mechanical difficulty or malfunction that occurred en route to or in a remote area;

(2) The pilot has satisfactorily completed an approved training program and is authorized in writing by the certificate holder for each item of preventive maintenance that the pilot is authorized to perform;

(3) There is no certificated mechanic available to perform preventive maintenance;

(4) The certificate holder has procedures to evaluate the accomplishment of a preventive maintenance item that requires a decision concerning the airworthiness of the rotorcraft; and

(5) The items of preventive maintenance authorized by this section are those listed in paragraph C of Appendix A of this Part.

(i) A manufacturer may-

(1) Rebuild or alter any aircraft, aircraft engine, propeller, or appliance manufactured by him under a type or production certificate;

(2) Rebuild or alter any appliance or part of aircraft, engines, propellers, or appliances manufactured by him under a Technical Standard Order Authorization, an FAA—Parts Manufacturer Approval, or Product and Process Specification issued by the Administrator; and

(3) Perform any inspection required by Part 91 or Part 125 of this chapter on aircraft it manufactures, while currently operating under a production certificate or under a currently approved production inspection system for such aircraft.

43.5 Approval for return to service after maintenance, preventive maintenance, rebuilding, or alteration

No person may approve for return to service any aircraft, airframe, aircraft engine, propeller, or appliance, that has undergone maintenance, preventive maintenance, rebuilding, or alteration unless-

(a) The maintenance record entry required by §43.9 or §43.1 1, as appropriate, has been made;

(b) The repair or alteration form authorized by or furnished by the Administrator has been executed in a manner prescribed by the Administrator; and

(c) If a repair or an alteration results in any change in the aircraft operating limitations or flight data contained in the approved aircraft flight manual, those operating limitations or flight data are appropriately revised and set forth as prescribed in §91.9 of this chapter.

43.7 Persons authorized to approve aircraft, airframes, aircraft engines, propellers, appliances, or component parts for return to service after maintenance, preventive maintenance, rebuilding, or alteration

(a) Except as provided in this section and §43.17, no person, other than the

Administrator, may approve an aircraft, airframe, aircraft engine, propeller, appliance, or component part for return to service after it has undergone maintenance, preventive maintenance, rebuilding, or alteration.

(b) The holder of a mechanic certificate or an inspection authorization may approve an aircraft, airframe, aircraft engine, propeller, appliance, or component part for return to service as provided in Part 65 of this chapter.

(c) The holder of a repair station certificate may approve an aircraft, airframe, aircraft engine, propeller, appliance, or component part for return to service as provided in Part 145 of this chapter.

(d) A manufacturer may approve for return to service any aircraft, airframe, aircraft engine, propeller, appliance, or component part which that manufacturer has worked on under §43.3(h). However, except for minor alterations, the work must have been done in accordance with technical data approved by the Administrator.

(e) The holder of an air carrier operating certificate or an operating certificate issued under Part 121, 127, or 135, may approve an aircraft, airframe, aircraft engine, propeller, appliance, or component part for return to service as provided in Part 121, 127, or 135 of this chapter, as applicable.

(f) A person holding at least a private pilot certificate may approve an aircraft for return to service after performing preventive maintenance under the provisions of §43.3(g).

43.9 Content, form, and disposition of maintenance, preventive maintenance,

rebuilding, and alteration records (except inspections performed in accordance with Part 91, Part 123, Part 125, §135.41 1 (a)(l), and §135.419 of this chapter)

(a) Maintenance record entries. Except as provided in paragraphs (b) and (c) of this section, each person who maintains, performs preventive maintenance, rebuilds, or alters an aircraft, airframe, aircraft engine, propeller, appliance, or component part shall make an entry in the maintenance record of that equipment containing the following information:

(1) A description (or reference to data acceptable to the Administrator) of work performed.
(2) The date of completion of the work performed.
(3) The name of the person performing the work if other than the person specified in paragraph (a) (4) of this section.
(4) If the work performed on the aircraft, airframe, aircraft engine, propeller, appliance, or component part has been performed satisfactorily, the signature, certificate number, and kind of certificate held by the person approving the work. The signature constitutes the approval for return to service only for the work performed. (See Fig. 8-1)

In addition to the entry required by this paragraph, major repairs and major alterations shall be entered on a form, and the form disposed of, in the manner prescribed in Appendix B, by the person performing the work.

ADVISORY CIRCULAR 43.9B—MAINTENANCE RECORDS

U.S. Department of Transportation
Federal Aviation Administration

Advisory Circular

Subject: MAINTENANCE RECORDS

Date: 1/9/84
Initiated by: AWS-340

AC No: 43-9B
Change:

1. PURPOSE. This advisory circular (AC) discusses maintenance record requirements under Federal Aviation Regulations (FAR) Part 43, Sections 43.9, 43.11, Part 91, Section 91.173, and the related responsibilities of owners, operators, and persons performing maintenance, preventive maintenance, and alterations.

2. CANCELLATION. AC 43-9A, Maintenance Records: General Aviation Aircraft, dated September 9, 1977, is cancelled.

3. RELATED FAR'S. FAR Parts 1, 43, 91, and 145.

4. BACKGROUND. The maintenance record requirements of Parts 43 and 91 have remained essentially the same for several years. Certain areas, however, continue to be misunderstood and changes to both Parts have recently been made. Those misunderstood areas and the recent changes necessitate this reiteration of Federal Aviation Administration (FAA) policy and an explanation of the changes.

5. DISCUSSION. Proper management of an aircraft operation begins with, and depends upon, a good maintenance record system. Properly executed and retained records provide owners, operators, and maintenance persons information essential in: controlling scheduled and unscheduled maintenance; evaluating the quality of maintenance sources; evaluating the economics and procedures of maintenance programs; troubleshooting; and eliminating the need for reinspection and/or rework to establish airworthiness. Only that information required to be a part of the maintenance record should be included and retained. Voluminous and irrelevant entries reduce the value of records in meeting their purposes.

6. MAINTENANCE RECORD REQUIREMENTS.

a. Responsibilities. Aircraft maintenance record keeping is a responsibility shared by the owner/operator and maintenance persons, with the ultimate responsibility assigned to the owner/operator by FAR Part 91, Section 91.165. Sections 91.165 and 91.173 set forth the requirements for owners and operators, while FAR Part 43, Sections 43.9 and 43.11 contain the requirements for maintenance persons. In general, the requirements for owners/operators and maintenance persons are the same; however, some small differences exist. These differences are discussed in this AC under the rule in which they exist.

b. Maintenance Record Entries Required. Section 91.165 requires each owner or operator to ensure that maintenance persons make appropriate entries in the maintenance records to indicate the aircraft has been approved

8-1 *Maintenance records.*

for return to service. Thus, the prime responsibility for maintenance records lies with the owner or operator. Section 43.9(a) requires persons performing maintenance, preventive maintenance, rebuilding, or alteration to make entries in the maintenance record of the equipment worked on. Maintenance persons, therefore, share the responsibility for maintenance records.

c. Maintenance Records are to be Retained. Section 91.173(a) sets forth the minimum content requirements and retention requirements for maintenance records. Maintenance records may be kept in any format which provides record continuity, includes required contents, lends itself to the addition of new entries, provides for signature entry, and is not confusing. Section 91.173(b) requires records of maintenance, alteration, and required or approved inspections to be retained until the work is repeated, superseded by other work, or for 1 year. It also requires the records, specified in Section 91.173(a)(2), to be retained and transferred with the aircraft at the time of sale.

NOTE: Section 91.173(a) contains an exception regarding work accomplished in accordance with Section 91.171. This does not exclude the making of entries for this work, but applies to the retention period of the records for work done in accordance with this section. The exclusion is necessary since the retention period of 1 year is inconsistent with the 24-month interval of test and inspection specified in Section 91.171. Entries for work done per this section are to be retained for 24 months or until the work is repeated or superseded.

d. Section 91.173(a)(1). This section requires a record of maintenance, for each aircraft (including the airframe) and each engine, propeller, rotor, and appliance of an aircraft. This does not require separate or individual records for each of these items. It does require the information specified in Sections 91.173(a)(1) through 91.173(a)(2)(vi) to be kept for each item as appropriate. As a practical matter, many owners and operators find it advantageous to keep separate or individual records since it facilitates transfer of the record with the item when ownership changes. Section 91.173(a)(1) has no counterpart in Section 43.9 or Section 43.11.

e. Section 91.173(a)(1)(i). This section requires the maintenance record entry to include "a description of the work performed." The description should be in sufficient detail to permit a person unfamiliar with the work to understand what was done, and the methods and procedures used in doing it. When the work is extensive, this results in a voluminous record and involves considerable time. To provide for this contingency, the rule permits reference to technical data acceptable to the Administrator in lieu of making the detailed entry. Manufacturer's manuals, service letters, bulletins, work orders, FAA advisory circulars, Major Repair and Alteration Forms (FAA Form 337), and others, which accurately describe what was done, or how it was done, may be referenced. Except for the documents mentioned, which are in common usage, referenced documents are to be made a part of the maintenance records and retained in accordance with Section 91.173(b). Certificated repair stations frequently work on components shipped to them when, as a consequence, the maintenance records are unavailable. To provide for this situation, repair stations should supply owners and operators with copies of work orders written for the work, in lieu of maintenance record entries. The work order copy must include the information required by Section 91.173(a)(1) through Section 91.173(a)(1)(iii), be made a part of the maintenance record, and retained per Section 91.173(b). This procedure is not the same as that for maintenance releases discussed in paragraph 17 of this AC, and it may not be used when maintenance records are available. Section 91.173(a)(1)(i) is identical to its counterpart, Section 43.9(a)(1), which imposes the same requirements on maintenance persons.

f. Section 91.173(a)(1)(ii) is identical to Section 43.9(a)(2) and requires entries to contain the date the work accomplished was completed. This is normally the date upon which the work is approved for return to service. However, when work is accomplished by one person and approved for return to service by another, the dates may differ. Two signatures may also appear under this circumstance; however, a single entry in accordance with Section 43.9(a)(3) is acceptable.

g. Section 91.173(a)(1)(iii) differs slightly from Section 43.9(a)(4) in that it requires the entry to indicate only the signature and certificate number of the person approving the work for return to service, and does not require the type of certificate being exercised to be indicated as does Section 43.9(a)(4). This is a new requirement of Section 43.9(a)(4), which assists owners and operators in meeting their responsibilities. Maintenance persons may indicate the type of certificate exercised by using A, P, A&P, IA, or RS for mechanic with airframe rating, with powerplant rating, with both ratings, with inspection authorization, or repair station, respectively.

8-1 *(Continued.)*

h. Section 91.173(a)(2) requires six items to be made a part of the maintenance record and maintained as such. Section 43.9 does not require maintenance persons to enter these items. Section 43.11 requires some of them to be part of entries made for inspections, but they are all the responsibility of the owner of operator. The six items are discussed as follows:

(1) Section 91.173(a)(2)(i) requires a record of total time in service to be kept for the airframe, each engine, and each propeller. FAR Part 1, Section 1.1, Definitions, defines "time in service," with respect to maintenance time records, as that time from the moment an aircraft leaves the surface of the earth until it touches it at the next point of landing. Section 43.9 does not require this to be part of the entries for maintenance, preventive maintenance, rebuilding, or alterations. However, Section 43.11 requires maintenance persons to make it a part of the entries for inspections made under Parts 91, 125, and Sections 135.411(a)(1) and 135.419. It is good practice to include time in service in all entries.

(i) Some circumstances impact the owner's or operator's ability to comply with Section 91.173(a)(2)(i). For example: in the case of rebuilt engines, the owner or operator would not have a way of knowing the "total time in service," since Section 91.175 permits the maintenance record to be discontinued and the engine time to be started at "zero." In this case, the maintenance record and the "time in service," subsequent to the rebuild, comprise a satisfactory record.

(ii) Many components, presently in service, were put into service prior to the requirements to keep maintenance records on them. Propellers are probably foremost in this group. In these instances, practicable procedures for compliance with the record requirements must be used. For example: "total time in service" may be derived using the procedures described in paragraph 13, Lost or Destroyed Records, of this AC; or if records prior to the regulatory requirements are just not available from any source, "time in service" may be kept since last complete overhaul. Neither of these procedures is acceptable when life-limited-parts status is involved or when AD compliance is a factor. Only the actual record since new may be used in these instances.

(iii) Sometimes engines are assembled from modules (turbojet and some turbopropeller engines) and a true "total time in service" for the total engine is not kept. If owners and operators wish to take advantage of this modular design, then "total time in service" and a maintenance record for each module is to be maintained. The maintenance records specified in Section 91.173(a)(2) are to be kept with the module.

(2) Section 91.173(a)(2)(ii) requires the current status of life-limited parts to be part of the maintenance record. If "total time in service" of the aircraft, engine, propeller, etc., is entered in the record when a life-limited part is installed and the "time in service" of the life-limited part is included, the normal record of time in service automatically meets this requirement.

(3) Section 91.173(a)(2)(iii) requires the maintenance record to indicate the time since last overhaul of all items installed on the aircraft which are required to be overhauled on a specified time basis. The explanation in paragraph 6h(2) of this AC also applies to this requirement.

(4) Section 91.173(a)(2)(iv) deals with the current inspection status and requires it to be reflected in the maintenance record. Again, the explanation in paragraph 6h(2) is appropriate even though Section 43.11(a)(2) requires maintenance persons to determine "time in service" of the item being inspected and to include it as part of the inspection entry.

(5) Section 91.173(a)(2)(v) requires the current status of applicable airworthiness directives (AD) to be a part of the maintenance record. The record is to include, at minimum, the method used to comply with the AD, the AD number, and revision date; and if the AD has requirements for recurring action, the time in service and the date when that action is required. When AD's are accomplished, maintenance persons are required to include the items specified in Section 43.9(a)(2), (3), and (4) in addition to those required by Section 91.173(a)(2)(v). An example of a maintenance record format for AD compliance is contained in Appendix 1 of this AC.

(6) Section 91.173(a)(2)(vi). In the past, the owner or operator has been permitted to maintain a list of current major alterations to the airframe, engine(s), propeller(s), rotor(s), or appliances. This procedure did not

produce a record of value to the owner/operator or to maintenance persons in determining the continued airworthiness of the alteration since such a record was not sufficient detail. This section of the rule has now been changed. It now prescribes that copies of the FAA Form 337, issued for the alteration, be made a part of the maintenance record.

7. PREVENTIVE MAINTENANCE.

a. Preventive maintenance is defined in Part 1, Section 1.1. Part 43, Appendix A, paragraph (c) lists those items which a pilot may accomplish under Section 43.3(g). Section 43.7 authorizes appropriately rated repair stations and mechanics, and persons holding at least a private pilot certificate to approve an aircraft for return to service after they have performed preventive maintenance. All of these persons must record preventive maintenance accomplished in accordance with the requirements of Section 43.9. Advisory Circular (AC)43-12, Preventive Maintenance, (as revised) contains further information on this subject.

b. The type of certificate exercised when maintenance or preventive maintenance is accomplished must be indicated in the maintenance record. Pilots may use PP, CP, or ATP to indicate private, commercial, or airline transport pilot certificate, respectively in approving preventive maintenance for return to service. Pilots are not authorized by Section 43.3(g) to perform preventive maintenance on aircraft when they are operated under Parts 121, 127, 129, or 135. Pilots may only approve for return to service preventive maintenance which they themselves have accomplished.

8. REBUILT ENGINE MAINTENANCE RECORDS.

a. Section 91.175 provides that "zero time" may be granted to an engine that has been rebuilt by a manufacturer or an agency approved by the manufacturer. When this is done, the owner/operator may use a new maintenance record without regard to previous operating history.

b. The manufacturer or an agency approved by the manufacturer that rebuilds and grants zero time to an engine is required by Section 91.175 to provide a signed statement containing: (1) the date the engine was rebuilt; (2) each change made as required by an AD; and (3) each change made in compliance with service bulletins, when the service bulletin specifically requests an entry to be made.

c. Section 43.2(b) prohibits the use of the term "rebuilt" in describing work accomplished in required maintenance records or forms unless the component worked on has had specific work functions accomplished. These functions are listed in Section 43.2(b) and, except for testing requirements, are the same as those set forth in Section 91.175. When terms such as "remanufactured," "reconditioned", or other terms coined by various aviation enterprises are used in maintenance records, owners and operators cannot assume that the functions outlined in Section 43.2(b) have been done.

9. RECORDING TACHOMETERS.

a. Time-in-service recording devices. These devices sense such things as: electrical power on, oil pressure, wheels on the ground, etc., and from these conditions provide an indication of time in service. With the exception of those which sense aircraft liftoff and touchdown, the indications are approximate.

b. Some owners and operators mistakenly believe these devices may be used in lieu of keeping time in service in the maintenance record. While they are of great assistance in arriving at time in service, such instruments, alone, do not meet the requirements of Section 91.173. For example, when the device fails and requires change, it is necessary to enter time in service and the instrument reading at the change. Otherwise, record continuity is lost.

10. MAINTENANCE RECORDS FOR AD COMPLIANCE. This subject is covered in AC 39-7, Airworthiness Directives for General Aviation Aircraft, as revised. A separate record may be kept for the airframe and each engine, propeller, rotor, and appliance, but is not required. This would facilitate record searches when inspection is needed, and when an engine, propeller, rotor, or appliance is removed, the record may be transferred with it. Such records may also be used as a schedule for recurring inspections. The format, shown in Appendix 1, is a suggested one, and adherence is not mandatory. Owners should be aware that they may be responsible for noncompliance with AD's

8-1 *(Continued.)*

when their aircraft are leased to foreign operators. They should, therefore, ensure that AD's are passed on to all their foreign lessees, and leases should be drafted to deal with this subject.

11. MAINTENANCE RECORDS FOR REQUIRED INSPECTIONS.

a. Section 43.11 contains the requirements for inspection entries. While these requirements are imposed on maintenance persons, owners and operators should become familiar with them in order to meet their responsibilities under Section 91.165.

b. The maintenance record requirements of Section 43.11 apply to the 100-hour, annual, and progressive inspections under Part 91; continuous inspection programs under Parts 91 and 125; Approved Airplane Inspection Programs under Part 135; and the 100-hour and annual inspections under Section 135.411(a)(1).

c. Misunderstandings persist regarding entry requirements for inspections under Section 91.169(e) (formerly Section 91.217). These requirements were formerly found in Section 43.9(a) and this contributed to misunderstanding. Appropriately rated mechanics without an inspection authorization (IA) are authorized to conduct these inspections and make the required entries. Particular attention should be given to Section 43.11(a)(7) in that it now requires a more specific statement than that previously required under Section 43.9. The entry, in addition to other items, must identify the inspection program used; identify the portion or segment of the inspection program accomplished; and contain a statement that the inspection was performed in accordance with the instructions and procedures for that program.

d. Questions continue regarding multiple entries for 100-hour/annual inspections. As discussed in paragraph 6d of this AC, neither Part 43 nor Part 91 requires separate records to be kept. Section 43.11, however, requires persons approving or disapproving equipment for return to service, after any required inspection, to make an entry in the record of that equipment. Therefore, when an owner maintains a single record, the entry of the 100-hour or annual inspection is made in that record. If the owner maintains separate records for the airframe, powerplants, and propellers, the entry for the 100-hour or annual inspection is entered in each.

12. DISCREPANCY LISTS.

a. Prior to October 15, 1982, issuance of discrepancy lists (or lists of defects) to owners or operators was appropriate only in connection with annual inspections under Part 91; inspections under Section 135.411(a)(1) of Part 135; continuous inspection programs under Part 125; and inspections under Section 91.217 of Part 91. Now, Section 43.11 requires that a discrepancy list be prepared by a person performing any inspection required by Parts 91, 125, or Section 135.411(a)(1) of Part 135.

b. When a discrepancy list is provided to an owner or operator, it says in effect, "except for these discrepancies, the item inspected is airworthy." It is imperative, therefore, that inspections be complete and that all discrepancies appear in the list. When circumstances dictate that an inspection be terminated before it is completed, the maintenance record should clearly indicate that the inspection was discontinued. The entry should meet all the other requirements of Section 43.11.

c. It is no longer a requirement that copies of discrepancy lists be forwarded to the local FAA District Office.

d. Discrepancy lists (or lists of defects) are part of the maintenance record and the owner/operator is responsible to maintain that record in accordance with Section 91.173(b)(3). The entry made by maintenance persons in the maintenance record should reference the discrepancy list when a list is issued.

13. LOST OR DESTROYED RECORDS. Occasionally, the records for an aircraft are lost or destroyed. This can create a considerable problem in reconstructing the aircraft records. First, it is necessary to reestablish the total time in service of the airframe. This can be done by: reference to other records which reflect the time in service; research of records maintained by repair facilities; and reference to records maintained by individual mechanics, etc. When these things have been done and the record is still incomplete, the owner/operator may make a notarized statement

1/9/84 AC 43-9B

in the new record describing the loss and establishing the time in service based on the research and the best estimate of time in service.

a. The current status of applicable AD's may present a more formidable problem. This may require a detailed inspection by maintenance personnel to establish that the applicable AD's have been complied with. It can readily be seen that this could entail considerable time, expense and, in some instances, might require recompliance with the AD.

b. Other items required by Section 91.173(a)(2), such as the current status of life-limited parts, time since last overhaul, current inspection status, and current list of major alterations, may present difficult problems. Some items may be easier to reestablish than others, but all are problems. Losing maintenance records can be troublesome, costly, and time consuming. Safekeeping of the records is an integral part of a good record system.

14. COMPUTERIZED RECORDS. There is a growing trend toward computerized maintenance records. Many of these systems are offered to owners/operators on a commercial basis. While these are excellent scheduling systems, alone, they normally to not meet the requirements of Sections 43.9 or 91.173. The owner/operator who uses such a system is required to ensure that it provides the information required by Section 91.173, including signatures. If not, modification to make them complete is the owners/operators responsibility and the responsibility may not be delegated.

15. PUBLIC AIRCRAFT. Prospective purchasers of aircraft, that have been used as public aircraft, should be aware that public aircraft are not subject to the certification and maintenance requirements in the FAR's and may not have records which meet the requirements of Section 91.173. Considerable research may be involved in establishing the required records when these aircraft are purchased and brought into civil aviation. The aircraft may not be certificated or used without such records.

16. LIFE-LIMITED PARTS.

a. Present day aircraft and powerplants commonly have life-limited parts installed. These life limits may be referred to as retirement times, service life limitations, parts retirement limitations, retirement life limits, life limitations, or other such terminology and may be expressed in hours, cycles of operation, or calendar time. They are set forth in type certificate data sheets, AD's, and operator's operations specifications, FAA-approved maintenance programs, the limitations section of FAA-approved airplane or rotorcraft flight manuals, and manuals required by operating rules.

b. Section 91.173(a)(2)(ii) requires the owner or operator of an aircraft with such parts installed to have records containing the current status of these parts. Many owners/operators have found it advantageous to have a separate record for such parts showing the name of the part, part number, serial number, date of installation, total time in service, date removed, and signature and certificate number of the person installing or removing the part. A separate record, as described, facilitates transferring the record with the part in the event the part is removed and later reinstalled or installed on another aircraft or engine. If a separate record is not kept, the aircraft record must contain sufficient information to clearly establish the status of the life-limited parts installed.

17. MAINTENANCE RELEASE.

a. In addition to those requirements discussed previously in this AC, Section 43.9 requires that major repairs and alterations be recorded as indicated in Appendix B of Part 43, (i.e., on FAA Form 337). An exception is provided in paragraph (b) of that appendix which allows repair stations certificated under Part 145 to use a maintenance release in lieu of the form for major repairs (and only major repairs).

b. The maintenance release must contain the information specified in paragraph (b)(3) of Appendix B of Part 43, be made a part of the aircraft maintenance record, and retained by the owner/operator as specified in Section 91.173. The maintenance release is usually a special document (normally a tag) and is attached to the product when it is approved for return to service. The maintenance release may, however, be on a copy of the work order written for the product. When this is done (it may be used only for major repairs) the entry on the work order must meet paragraph (b)(3) of the appendix.

8-1 *(Continued.)*

c. Some repair stations use what they call a maintenance release for other than major repairs. This is sometimes a tag and sometimes information on a work order. When this is done, all of the requirements of Section 43.9 must be met (those of (b)(3) of the appendix are not applicable) and the document is to be made and retained as part of the maintenance records under Section 91.173. This was discussed in paragraph 6e of this AC.

18. FAA FORM 337.

a. Major repairs and alterations are to be recorded on FAA Form 337, Major Repair and Alteration, as stated in paragraph 17. This form is executed by the person making the repair or alteration. Provisions are made on the form for a person other than that person performing the work to approve the repair or alteration for return to service.

b. These forms are now required to be made part of the maintenance record of the product repaired or altered and retained in accordance with Section 91.173.

c. Detailed instructions for use of this form are contained in AC 43.9-1D, Instructions for Completion of FAA Form 337, Major Repair and Alteration.

d. Some manufacturers have initiated a policy of indicating, on their service letters and bulletins, and other documents dealing with changes to their aircraft, whether or not the changes constitute major repairs or alterations. They also indicate when, in their opinion, a Form 337 lies with the person accomplishing the repairs or alterations and cannot be delegated. When there is a question, it is advisable to contact the local office of the FAA for guidance.

19. TESTS AND INSPECTIONS FOR ALTIMETER SYSTEMS, ALTITUDE REPORTING EQUIPMENT, AND ATC TRANSPONDERS. The recordation requirements for these tests and inspections are the same as for other maintenance. There are essentially three tests and inspections, (the altimeter system, the transponder system, and the data correspondence test) each of which may be subdivided relative to who may perform specific portions of the test. The basic authorization for performing these tests and inspections, found in Section 43.3, are supplemented by Sections 91.171 and 91.172. When multiple persons are involved in the performance of tests and inspections, care must be exercised to insure proper authorization under these three sections and compliance with Sections 43.9 and 43.9(a)(3) in particular.

20. BEFORE YOU BUY. This is the proper time to take a close look at the maintenance records of any used aircraft you expect to purchase. A well-kept set of maintenance records, which properly identifies all previously performed maintenance, alterations, and AD compliances, is generally a good indicator of the aircraft condition. This is not always the case, but in any event, before you buy, require the owner to produce the maintenance records for your examination, and require correction of any discrepancies found on the aircraft or in the records. Many prospective owners have found it advantageous to have a reliable unbiased maintenance person examine the maintenance records, as well as the aircraft, before negotiations have progressed too far. If the aircraft is purchased, take the time to review and learn the system of the previous owner to ensure compliance and continuity when you modify or continue that system.

M. C. Beard
Director of Airworthiness

(b) Each holder of an air carrier operating certificate or an operating certificate issued under Part 121, 127, or 135, that is required by its approved operations specifications to provide for a continuous airworthiness maintenance program, shall make a record of the maintenance, preventive maintenance, rebuilding, and alteration, on aircraft, airframes, aircraft engines, propellers, appliances, or component parts which it operates in accordance with the applicable provisions of Part 121, 127, or 135 of this chapter, as appropriate.

(c) This section does not apply to persons performing inspections in accordance with Part 91, 123, 125, §135.41 l(a) (1), or §135.419 of this chapter.

43.11 Content, form, and disposition of records for inspections conducted under Parts 91, and 125 and §§135.41 1 (a) (1) and 135.419 of this chapter

(a) Maintenance record entries. The person approving or disapproving for return to service an aircraft, airframe, aircraft engine, propeller, appliance, or component part after any inspection performed in accordance with Part 91, 123, 125, §135.41 1 (a) (1), or §135.419 shall make an entry in the maintenance record of that equipment containing the following information:

(1) The type of inspection and a brief description of the extent of the inspection.
(2) The date of the inspection and aircraft total time in service.
(3) The signature, the certificate number, and kind of certificate held by the person approving or disapproving for

return to service the aircraft, airframe, aircraft engine, propeller, appliance, component part, or portions thereof.

(4) Except for progressive inspections, if the aircraft is found to be airworthy and approved for return to service, the following or a similarly worded statement—"I certify that this aircraft has been inspected in accordance with (insert type) inspection and was determined to be in airworthy condition."

(5) Except for progressive inspections, if the aircraft is not approved for return to service because of needed maintenance, noncompliance with applicable specifications, airworthiness directives, or other approved data, the following or a similarly worded statement—"I certify that this aircraft has been inspected in accordance with (insert type) inspection and a list of discrepancies and unairworthy items dated (date) has been provided for the aircraft owner or operator." (See Fig. 8-2.)

(6) For progressive inspections, the following or a similarly worded statement—"I certify that in accordance with a progressive inspection program, a routine inspection of (identify whether aircraft or components) and a detailed inspection of (identify components) were performed and the (aircraft or components) are (approved or disapproved) for return to service." If disapproved, the entry will further state "and a list of discrepancies and unairworthy items dated (date) has been provided to the aircraft owner or operator."

ADVISORY CIRCULAR 39-7B—AIRWORTHINESS DIRECTIVES

U.S. Department of Transportation
Federal Aviation Administration

Advisory Circular

Subject: AIRWORTHINESS DIRECTIVES

Date: 4/8/87
Initiated by: AFS-340

AC No: 39-7B
Change:

1. PURPOSE. This advisory circular (AC) provides guidance and information to owners and operators of aircraft concerning their responsibility for complying with airworthiness directives (AD's) and recording AD compliance in the appropriate maintenance records.

2. CANCELLATION. AC 39-7A, Airworthiness Directives for General Aviation Aircraft, dated September 17, 1982, is cancelled.

3. RELATE FEDERAL AVIATION REGULATIONS (FAR). FAR Part 39; FAR Part 43, Sections 43.9 and 43.11; FAR Part 91, Sections 91.163, 91.165, and 91.173.

4. BACKGROUND. Title VI of the Federal Aviation Act of 1958, as amended by Section 6 of the Department of Transportation Act, defines the Federal Aviation Administration (FAA) role regarding the promotion of safety of flight for civil aircraft. One safety function charged to the FAA is to require correction of unsafe conditions discovered in any product (aircraft, aircraft engine, propeller, or appliance) after type certification or other approval, when that condition is likely to exist or develop in other products of the same type design. AD's are used by the FAA to notify aircraft owners and operators of unsafe conditions and to require their correction. Ad's prescribe the conditions and limitations, including inspections, repair, or alteration under which the product may continue to be operated. AD's are FAR codified in FAR Part 39 and issued in accordance with the public rulemaking procedures of the Administrative Procedure Act, Title 5, U.S.C. Section 553.

5. AD CATEGORIES. Since AD's are FAR, they are published in the Federal Register as amendments to FAR Part 39. Depending on the urgency, AD's are issued as follows:

a. Normally a notice of proposed rulemaking (NPRM) for an AD is issued and published in the Federal Register when an unsafe condition is believed to exist in a product. Interested persons are invited to comment on the NPRM by submitting such written data, views, or arguments as they may desire. The comment period is usually 30 days. Proposals contained in the notice may be changed or withdrawn in light of comments received. When the final rule resulting from the NPRM is adopted, it is published in the Federal Register, printed, and distributed by first class mail to the registered owners of the products affected.

b. Emergency AD's. AD's of an urgent nature are adopted without prior notice (NPRM) under emergency procedures as immediate adopted rules. The AD's normally become effective in less than 30 days after publication

8-2 *Airworthiness directives.*

in the Federal Register and are distributed by telegram or first class mail to the registered owners of the product affected.

6. AD'S WHICH APPLY TO OTHER THAN AIRCRAFT. AD's may be issued which apply to engines, propellers, or appliances installed on multiple makes or models of aircraft. When the product can be identified as being installed on a specific make or model aircraft, AD distribution is made to the registered owners of those aircraft. However, there are times when a determination cannot be made, and direct distribution to the registered owner is impossible. For this reason, aircraft owners and operators are urged to subscribe to the Summary of Airworthiness Directives which contains all previously published AD's and a biweekly supplemental service. The Summary of Airworthiness Directives is sold and distributed for the Superintendent of Documents by the FAA in Oklahoma City. AC 39-6L, Announcement of Availability – Summary of Airworthiness Directives, provides ordering information and subscription prices on these publications. AC 39-6L may be obtained, without cost, from the U.S. Department of Transportation, Utilization and Storage Section, M-443.2, Washington, D.C. 20590.

7. APPLICABILITY OF AD'S. Each AD contains an applicability statement specifying the product (aircraft, aircraft engine, propeller, or appliance) to which it applies. Some aircraft owners and operators mistakenly assume that AD's are not applicable to aircraft with experimental or restricted airworthiness certificates. Unless specifically limited, AD's apply to the make and model set forth in the applicability statement regardless of the kind of airworthiness certificate issued for the aircraft. Type certificate and airworthiness certification information are used to identify the product affected. When there is no reference to serial numbers, all serial numbers are affected. Limitations may be placed on applicability by specifying the serial number or number series to which the AD is applicable. The following are examples of AD applicability statements:

a. "Applies to Robin RA-15-150 airplanes." This statement makes the AD applicable to all airplanes of the model listed, regardless of type of airworthiness certificate issued to the aircraft and includes standard, restricted, limited, or experimental airworthiness certificates.

b. "Applies to Robin RA-15-150 airplanes except those certificated in the restricted category." This statement, or one similarly worded, incorporates all airplanes of the model listed, except those in the restricted category and is applicable to experimental aircraft.

c. "Applies to Robin RA-15-150 airplanes certificated in all categories excluding experimental aircraft." This statement incorporates all airplanes including restricted category of the model listed except those issued experimental certificates.

8. AD COMPLIANCE. AD's are regulations issued under FAR Part 39. Therefore, no person may operate a product to which an AD applies, except in accordance with the provisions of the AD. It should be understood that to "operate" not only means piloting the aircraft, but also causing or authorizing the product to be used. Compliance with emergency AD's can be a problem for operators of leased aircraft. The FAA has no means available for making notification to other than registered owners. Therefore, it is important that owners of leased aircraft make the AD information available to the operators leasing their aircraft as expeditiously as possible. Unless this is done, the lessee may not be aware of the AD and safety may be jeopardized.

9. COMPLIANCE TIME OR DATE.

a. The belief that AD compliance is only required at the time of a required inspection, e.g., at 100 hours of annual inspection is not correct. The required compliance time is specified in each AD, and no person may operate the affected product after expiration of that stated compliance time without an exemption or a special flight authorization when the AD specifically permits such operation.

b. Compliance requirements specified in AD's are established for safety reasons and may be stated in numerous ways. Some AD's are of such a serious nature they require compliance before further flight. In some instances the AD authorizes flight, provided a ferry permit is obtained, but without such authorization in the AD, further flight is prohibited. Other AD's express compliance time in terms of a specific number of hours of operation, for example, "compliance required within the next 50-hours time in service after the effective date of this Ad." Compliance times may also be expressed in operational terms such as, "within the next 10 landings after the effective date of this AD." For turbine engines, compliance times are often expressed in terms of cycles. A cycle normally consists of an engine start, takeoff operation, landing, and engine shutdown. When a direct relationship between airworthiness and calendar time is identified, compliance time may be expressed as a calendar date. Another aspect of compliance times to be emphasized is that not all AD's have a one-time compliance. Repetitive inspections at specified intervals after initial compliance may be required. Repetitive inspection is used in lieu of a fix because of costs or until a fix is developed.

10. ADJUSTMENTS IN COMPLIANCE REQUIREMENTS. In some instances, a compliance time other than that specified in the AD would be advantageous to the owner/operator. In recognition of this need, and when equivalent safety can be shown, flexibility is provided by a statement in the AD allowing adjustment of the specified interval. When adjustment authority is provided in an AD, owners or operators desiring to make an adjustment are required to submit data substantiating their proposed adjustment to their FAA district office for consideration. The person authorized to approve adjustments in compliance requirements is normally identified in the AD.

11. EQUIVALENT MEANS OF COMPLIANCE. Most AD's indicate the acceptability of an equivalent means of compliance. It cannot be assumed that only one specific repair, modification, or inspection method is acceptable to correct an unsafe condition; therefore, development of alternatives is not precluded. An equivalent means of compliance must be substantiated and "FAA approved." Normally the person authorized to approve an alternate method of compliance is indicated by title and address on the AD.

12. RESPONSIBILITY FOR AD COMPLIANCE AND RECORDATION. Responsibility for AD compliance always lies with the registered owner or operator of the aircraft.

a. This responsibility may be met by ensuring that certificated and appropriately rated maintenance persons accomplish the maintenance required by the AD and properly record it in the maintenance records. This must be accomplished within the compliance time specified in the AD or the aircraft may not be operated.

b. Maintenance persons may also have direct responsibility for AD compliance, aside from the times when AD compliance is the specific work contracted for by the owner/operator. When a 100-hour, annual, or progressive inspection, or an inspection required under FAR Parts 125 or 135 is accomplished, FAR Section 43.15(a) requires the person performing the inspection to perform it so that all applicable airworthiness requirements are met, which includes compliance with AD's.

c. Maintenance persons should note that even though an inspection of the complete aircraft is not made, if the inspection conducted is a Progressive Inspection, an inspection required by FAR Part 125 determination of AD compliance for those portions of the aircraft inspected is required.

d. For aircraft inspected in accordance with a continuous inspection program under FAR Part 91, Section 91.169(f), inspection persons are required to comply with AD's only when the portions of the inspection program provided to them require compliance. The program may require a determination of AD compliance for the entire aircraft by a general statement, or compliance with AD's applicable only to portions of the aircraft being inspected, or it may not require compliance at all. This does not mean AD compliance is not required at the compliance time or date specified in the AD. It only means that the owner or operator has elected to handle AD compliance apart from the inspection program. The owner or operator remains fully responsible for AD compliance.

e. The person accomplishing the AD is required by FAR Part 43, Section 43.9, to record AD compliance. The entry must include those items specified in FAR Section 43.9(a)(1) through (a)(4). The owner is required, by FAR Part 91, Section 91.165, to ensure that maintenance personnel make appropriate entries and, by FAR Section 91.173, to maintain those records. It should be noted that there is a difference between the records required to be kept under FAR Section 91.173 and those FAR Section 43.9 requires maintenance personnel to make. Owners and operators may add this required information themselves or request maintenance personnel to include it in the entry they make. In either case, the owner/operator is responsible for keeping proper records.

8-2 *(Continued.)*

4/8/87 AC 39-7E

f. Certain AD's permit pilots to perform checks of some items under specific conditions. The AD's normally include recording requirements which are the same as those specified in FAR Section 43.9. However, if the AD does not include recording requirements for the pilot, FAR Parts 43 and 91, Section 91.173(a)(1) and (a)(2), require the owner/operator to make and keep certain minimum records for specific times. The person who accomplished the work, who returned the aircraft to service, and the status of AD compliance are among these required records.

13. SOME AD'S REQUIRE REPETITIVE OR PERIODIC INSPECTION. In order to provide for flexibility in administering such AD's, an AD may provide for adjustment of the inspection interval to coincide with inspections required by FAR Part 91, or other regulations. The conditions under which this may be done and approval requirements are stated in the AD. If the AD does not contain such provision, adjustments are not permitted. However, amendment, modification, or adjustment of the terms of the AD may be requested by contacting the office which issued the AD or by the petition procedures provided in FAR Part 11.

14. SUMMARY. The registered owner or operator of the aircraft is responsible for compliance with AD's applicable to airframes, powerplants, propellers, appliances, and parts and components thereof for all aircraft they operate. Maintenance personnel are also responsible for AD compliance when they accomplish an inspection required by FAR Part 91.

William T. Brennan
Acting Director of Flight Standards

8-2 *(Continued.)*

(7) If an inspection is conducted under an inspection program provided for in Part 91, 123, 125, or §135.41 1 (a) (1), the entry must identify the inspection program, that part of the inspection program accomplished, and contain a statement that the inspection was performed in accordance with the inspections and procedures for that particular program.

(b) Listing of discrepancies and placards. If the person performing any inspection required by Part 91 or 125 or 135.41 1 (a)(l) of this chapter finds that the aircraft is unairworthy or does not meet the applicable type certificate data, airworthiness directives, or other approved data upon which its airworthiness depends, that person must give the owner or lessee a signed and dated list of those discrepancies. For those items permitted to be inoperative under §91.30(d)(2), that person shall place a placard, that meets the aircraft's airworthiness certification regulations, on each inoperative instrument and the cockpit control of each item of inoperative equipment, marking it "Inoperative," and shall add the items to the signed and dated list of discrepancies given to the owner or lessee.

43.12 Maintenance records: falsification, reproduction, or alteration

(a) No person may make or cause to be made:

(1) Any fraudulent or intentionally false entry in any record or report that is required to be made, kept, or used to show compliance with any requirement under this Part;

(2) Any reproduction, for fraudulent purpose, of any record or report under this Part;

(3) Any alteration, for fraudulent purpose, of any record or report under this Part.

(b) The commission by any person of an act prohibited under paragraph (a) of this section is a basis for suspending or revoking the applicable airman, operator, or production certificate, Technical Standard Order Authorization, FAA—Parts Manufacturer Approval, or Product and Process Specification issued by the Administrator and held by that person.

43.13 Performance rules (general)

(a) Each person performing maintenance alteration, or preventive maintenance on an aircraft, engine, propeller, or appliance shall use the methods, techniques, and practices prescribed in the current manufacturer's maintenance manual or Instructions for Continued Airworthiness prepared by its manufacturer, or other methods, techniques, and practices acceptable to the Administrator, except as noted in §43.16. He shall use the tools, equipment, and test apparatus necessary to assure completion of the work in accordance with accepted industry practices. If special equipment or test apparatus is recommended by the manufacturer involved, he must use that equipment or apparatus or its equivalent acceptable to the Administrator.

(b) Each person maintaining or altering, or performing preventive maintenance, shall do that work in such a manner and use materials of such a quality, that the condition of the aircraft, airframe, aircraft engine, propeller, or appliance worked on will be at least equal to its original or properly altered condition (with regard to aerodynamic function, structural strength, resistance to vibration and deterioration, and other qualities affecting airworthiness).

(c) Special provisions for holders of air carrier operating certificates and operating certificates issued under the provisions of Part 121, 127, or 135 and Part 129 operators holding operations specifications. Unless otherwise notified by the Administrator, the methods, techniques, and practices contained in the maintenance manual or the maintenance part of the manual of the holder of an air carrier operating certificate or an operating certificate under Part 121, 127, or 135 (that is required by its operating specifications to provide a continuous airworthiness maintenance and inspection program) and Part 129 operators holding operations specifications constitute acceptable means of compliance with this section.

43.15 Additional performance rules for inspections

(a) General. Each person performing an inspection required by part 91, 123, 125, or 135 of this chapter, shall—

(1) Perform the inspection so as to determine whether the aircraft, or portion(s) thereof under inspection, meets all applicable airworthiness requirements; and
(2) If the inspection is one provided for in Part 123, 125, 135, or §91.409(e) of this chapter, perform the inspection in accordance with the instructions and procedures set forth in the inspection program for the aircraft being inspected.

(b) Rotorcraft. Each person performing an inspection required by Part 91 on a rotorcraft shall inspect the following systems in accordance with the maintenance manual or Instructions for Continued Airworthiness of the manufacturer concerned:

(1) The drive shafts or similar systems.
(2) The main rotor transmission gear box for obvious defects.
(3) The main rotor and center section (or the equivalent area).
(4) The auxiliary rotor on helicopters.

(c) Annual and 100-hour inspections.

(1) Each person performing an annual or 100-hour inspection shall use a checklist while performing the inspection. The checklist may be of the person's own

design, one provided by the manufacturer of the equipment being inspected, or one obtained from another source. This checklist must include the scope and detail of the items contained in Appendix D to this Part and paragraph (b) of this section.

(2) Each person approving a reciprocating-engine-powered aircraft for return to service after an annual or 100-hour inspection shall, before that approval, run the aircraft engine or engines to determine satisfactory performance, in accordance with the manufacturer's recommendations, of—

(i) Power output (static and idle rpm);
(ii) Magnetos;
(iii) Fuel and oil pressure; and
(iv) Cylinder and oil temperature.

(3) Each person approving a turbine-engine-powered aircraft for return to service after an annual, 100-hour, progressive inspection shall, before that approval, run the aircraft engine or engines to determine satisfactory performance in accordance with the manufacturer's recommendations.

(d) Progressive inspection.

(1) Each person performing a progressive inspection shall, at the start of a progressive inspection system, inspect the aircraft completely. After this initial inspection, routine and detailed inspections must be conducted as prescribed in the progressive inspection schedule. Routine inspections consist of visual examination or check of the appliances,

the aircraft, and its components and systems, insofar as practicable without disassembly. Detailed inspections consist of a thorough examination of the appliances, the aircraft, and its components and systems, with such disassembly as is necessary. For the purposes of this subparagraph, the overhaul of a component or system is considered to be a detailed inspection. If the aircraft is away from the station where inspections are normally conducted, an appropriately rated mechanic, a certificated repair station, or the manufacturer of the aircraft may perform inspections in accordance with the procedures and using the forms of the person who would otherwise perform the inspection.

43.16 Airworthiness Limitations

Each person performing an inspection or other maintenance specified in an Airworthiness Limitations section of a manufacturer's maintenance manual or Instructions for Continued Airworthiness shall perform the inspection or other maintenance in accordance with that section, or in accordance with operations specifications approved by the Administrator under Parts 121, 123, 127, or 135, or an inspection program approved under §91.409(e).

43.17 Maintenance, preventive maintenance, and alterations performed on U.S. aeronautical products by certain Canadian persons

(a) Definitions. For purposes of this section:

Aeronautical product means any civil aircraft or airframe, aircraft engine, propeller, appliance, component, or part to be installed thereon.

Canadian aeronautical product means any civil aircraft or airframe, aircraft engine, propeller, or appliance under airworthiness regulations by the Canadian Department of Transport, or component or part to be installed thereon.

U.S. aeronautical product means any civil aircraft or airframe, aircraft engine, propeller, or appliance under airworthiness regulation by the FAA, or component or part to be installed thereon.

(b) Applicability. This section does not apply to any U.S. aeronautical products maintained or altered under any bilateral agreement made between Canada and any other than the United States.

(c) Authorized persons.

(1) A person holding a valid Canadian Department of Transport license (Aircraft Maintenance Engineer) and appropriate ratings may, with respect to a U.S.-registered aircraft located in Canada, perform maintenance, preventive maintenance, and alterations in accordance with the requirements of paragraph (d) of this section and approve the affected aircraft for return to service in accordance with the requirements of paragraph (e) of this section.

(2) A company (Approved Maintenance Organization) (AMO) whose system of quality control for the maintenance, alteration, and inspection of aeronautical products has been approved by the Canadian Department of Transport, or a

person who is an authorized employee performing work for such a company may, with respect to a U.S.-registered aircraft located in Canada or other U.S. aeronautical products transported to Canada from the United States, perform maintenance, preventive maintenance, and alterations in accordance with the requirements of paragraph (d) of this section and approve the affected products for return to service in accordance with the requirements of paragraph (e) of this section.

(d) Performance requirements. A person authorized in paragraph (c) of this section may perform maintenance (including any inspection required by §91.409 of this chapter except an annual inspection), preventive maintenance, and alterations, provided:

(1) The person performing the work is authorized by the Canadian Department of Transport to perform the same type of work with respect to Canadian aeronautical products.

(2) The work is performed in accordance with §§43.13, 43.15, 43.16 of this chapter, as applicable.

(3) The work is performed such that the affected product complies with the applicable requirements of Part 36 of this chapter, and

(4) The work is recorded in accordance with §§43.2(a), 43.9, and 43.11 of this chapter, as applicable.

(e) Approval requirements.

(1) To return an affected product to service, a person authorized in paragraph (c) of

this section must approve (certify) maintenance, preventive maintenance, and alterations performed under this section, except that an Aircraft Maintenance Engineer may not approve a major repair or major alteration.

(2) An AMO whose system of quality control for the maintenance, preventive maintenance, alteration, and inspection of aeronautical products has been approved by the Canadian Department of Transport, or an authorized employee performing work for such an AMO, may approve (certify) a major repair or major alteration performed in accordance with technical data approved by the Administrator.

(f) No person may operate in air commerce an aircraft, airframe, aircraft engine, propeller, or appliance on which maintenance, preventive maintenance, or alteration has been performed under this section unless it has been approved for return to service by a person authorized in this section.

Appendix A Major Alterations, Major Repairs, and Preventive Maintenance

(a) Major alterations.

(1) Air frame major alterations. Alterations of the following parts and alterations of the following types, when not listed in the aircraft specifications issued by the FAA, are airframe major alterations:

(i) Wings.
(ii) Tail surfaces.
(iii) Fuselage.
(iv) Engine mounts.
(v) Control system.

(vi) Landing gear.
(vii) Hull or floats.
(viii) Elements of an airframe including spars, ribs, fittings, shock absorbers, bracing, cowlings, fairings, and balance weights.
(ix) Hydraulic and electrical actuating system of components.
(x) Rotor blades.
(xi) Changes to the empty weight or empty balance which result in an increase in the maximum certificated weight or center of gravity limits of the aircraft.
(xii) Changes to the basic design of the fuel, oil, cooling, heating, cabin pressurization, electrical, hydraulic, de-icing, or exhaust systems.
(xiii) Changes to the wing or to fixed or movable control surfaces which affect flutter and vibration characteristics.

(2) Powerplant major alterations. The following alterations of a powerplant when not listed in the engine specifications issued by the FAA, are powerplant major alterations:

(i) Conversion of an aircraft engine from one approved model to another, involving any changes in compression ratio, propeller reduction gear, impeller gear ratios or the substitution of major engine parts which requires extensive rework and testing of the engine.
(ii) Changes to the engine by replacing aircraft engine structural parts with parts not supplied by the original

manufacturer or parts not specifically approved by the Administrator.

(iii) Installation of an accessory which is not approved for the engine.

(iv) Removal of accessories that are listed as required equipment on the aircraft or engine specification.

(v) Installation of structural parts other than the type of parts approved for the installation.

(vi) Conversions of any sort for the purpose of using fuel of a rating or grade other than that listed in the engine specifications.

(3) Propeller major alterations. The following alterations of a propeller when not authorized in the propeller specifications issued by the FAA are propeller major alterations:

(i) Changes in blade design.

(ii) Changes in hub design.

(iii) Changes in the governor or control design.

(iv) Installation of a propeller governor or feathering system.

(v) Installation of propeller de-icing system.

(vi) Installation of parts not approved for the propeller.

(4) Appliance major alterations. Alterations of the basic design not made in accordance with recommendations of the appliance manufacturer or in accordance with an FAA Airworthiness Directive are appliance major alterations. In addition, changes in the basic design of radio communication

and navigation equipment approved under type certification or a Technical Standard Order that have an effect on frequency stability, noise level, sensitivity, selectivity, distortion, spurious radiation, AVC characteristics, or ability to meet environmental test conditions and other changes that have an effect on the performance of the equipment are also major alterations.

(b) Major repairs.

(1) Air frame major repairs. Repairs to the following parts of an airframe and repairs of the following types, involving the strengthening, reinforcing, splicing, and manufacturing of primary structural members or their replacement, when replacement is by fabrication such as riveting or welding, are airframe major repairs:

(i) Box beams.
(ii) Monocoque or semimonocoque wings or control surfaces.
(iii) Wing stringers or chord members.
(iv) Spars.
(v) Spar flanges.
(vi) Members of truss-type beams.
(vii) Thin sheet webs of beams.
(viii) Keel and chine members of boat hulls or floats.
(ix) Corrugated sheet compression members which act as flange material of wings or tail surfaces.
(x) Wing main ribs and compression members.
(xi) Wing or tail surface brace struts.
(xii) Engine mounts.

(xiii) Fuselage longerons.
(xiv) Members of the side truss, horizontal truss, or bulkheads.
(xv) Main seat support braces and brackets.
(xvi) Landing gear brace struts.
(xvii) Axles.
(xviii) Wheels.
(xix) Skis, and ski pedestals.
(xx) Parts of the control system such as control columns, pedals, shafts, brackets, or horns.
(xxi) Repairs involving the substitution of material.
(xxii) The repair of damaged areas in metal or plywood stressed covering exceeding six inches in any direction.
(xxiii) The repair of portions of skin sheets by making additional seams.
(xxiv) The splicing of skin sheets.
(xxv) The repair of three or more adjacent wing or control surface ribs or the leading edge of wings and control surfaces, between such adjacent ribs.
(xxvi) Repair of fabric covering involving an area greater than that required to repair two adjacent ribs.
(xxvii) Replacement of fabric on fabric covered parts such as wings, fuselages, stabilizers, and control surfaces.
(xxviii) Repairing, including rebottoming, of removable or integral fuel tanks and oil tanks.

(2) Powerplant major repairs. Repairs of the following parts of an engine and repairs of the following types, are powerplant major repairs:
   (i) Separation or disassembly of a crankcase or crankshaft of a reciprocating engine equipped with an integral supercharger.
   (ii) Separation or disassembly of a crankcase or crankshaft of a reciprocating engine equipped with other than spurtype propeller reduction gearing.
   (iii) Special repairs to structural engine parts by welding, plating, metallizing, or other methods.

(3) Propeller major repairs. Repairs of the following types to a propeller are propeller major repairs:
   (i) Any repairs to, or straightening of steel blades.
   (ii) Repairing or machining of steel hubs.
   (iii) Shortening of blades.
   (iv) Retipping of wood propellers.
   (v) Replacement of outer laminations on fixed-pitch wood propellers.
   (vi) Repairing elongated bolt holes in the hub of fixed-pitch wood propellers.
   (vii) Inlay work on wood blades.
   (viii) Repairs to composition blades.
   (ix) Replacement of tip fabric.
   (x) Replacement of plastic covering.
   (xi) Repair of propeller governors.
   (xii) Overhaul of controllable pitch propellers.

(xiii) Repairs to deep dents, cuts, scars, nicks, etc., and straightening of aluminum blades.

(xiv) The repair or replacement of internal elements of blades.

(4) Appliance major repairs. Repairs of the following types to appliances are appliance mayor repairs:

(i) Calibration and repair of instruments.

(ii) Calibration of radio equipment.

(iii) Rewinding the field coil of an electrical accessory.

(iv) Complete disassembly of complex hydraulic power valves.

(v) Overhaul of pressure type carburetors, and pressure type fuel, oil, and hydraulic pumps.

Because numerous pilots and owners repaint their aircraft the following regulations regarding nationality and registration marks are important.

45.21 General

The following regulations regarding nationality and registration marks are important.

(a) Except as provided in §45.22, no person may operate a United States registered aircraft unless that aircraft displays nationality and registration marks in accordance with the requirements of this section and §§45.23 through 45.33.

(b) Unless otherwise authorized by the Administrator, no person may place on any aircraft a design, mark, or symbol that modifies or confuses the nationality and registration marks.

(c) Aircraft nationality and registration marks must—

(1) Except as provided in paragraph (d) of this section, be painted on the aircraft or affixed by any other means insuring a similar degree of permanence;
(2) Have no ornamentation;
(3) Contrast in color with the background and
(4) Be legible.

(d) The aircraft nationality and registration marks may be affixed to an aircraft with readily removable material if-

(1) It is intended for immediate delivery to a foreign purchaser;
(2) It is bearing a temporary registration number: or
(3) It is marked temporarily to meet the requirements of §45.22(c)(1) or §45.29(h) of this part, or both.

45.22 Exhibition, antique, and other aircraft: Special rules

(a) When display of aircraft nationality and registration marks in accordance with §§45.21 and 45.23 through 45.33 would be inconsistent with exhibition of that aircraft, a United States registered aircraft may be operated without displaying those marks anywhere on the aircraft if—

(1) It is operated for the purpose of exhibition, including a motion picture or television production, or an air show;
(2) Except for practice and test flights necessary for exhibition purposes, it is operated only at the location of the

exhibition, between the exhibition locations, and between those locations and the base of operations of the aircraft; and

(3) For each flight in the United States:

(i) It is operated with the prior approval of the Flight Standards District Office, in the case of a flight within the lateral boundaries of the surface areas of Class B, Class C, Class D, or Class E airspace designated for the takeoff airport, or within 4.4 nautical miles of that airport if it is within Class G airspace; or

(ii) It is operated under a flight plan filed under either §91.153 or §91.169 of this chapter describing the marks it displays, in the case of any other flight.

(b) A small U.S.-registered aircraft built at least 30 years ago or a U.S.-registered aircraft for which an experimental certificate has been issued under §21.191(d) or §21.191(g) for operation as an exhibition aircraft or as an amateur-built aircraft and which has the same external configuration as an aircraft built at least 30 years ago may be operated without displaying marks in accordance with §§45.21 and 45.23 through 45.33 if—

(1) It displays in accordance with §45.21 (c) marks at least 2 inches high on each side of the fuselage or vertical tail surface consisting of the Roman capital letter "N" followed by-

(i) The U.S. registration number of the aircraft; or

(ii) The symbol appropriate to the airworthiness certificate of the aircraft

("C," standard, "R," restricted, "L," limited; or "X," experimental) followed by the U.S. registration number of the aircraft and

(2) It displays no other mark that begins with the letter "N" anywhere on the aircraft, unless it is the same mark that is displayed under subparagraph (1) of this paragraph.

(c) No person may operate an aircraft under paragraph (a) or (b) of this section—

(1) In an ADIZ or DEWIZ described in Part 99 of this chapter unless it temporarily bears marks in accordance with §§45.21 and 45.23 through 45.3 3;

(2) In a foreign country unless that country consents to that operation; or

(3) In any operation conducted under Part 121, 127, 133, 135, or 137 of this chapter.

(d) If, due to the configuration of an aircraft, it is impossible for a person to mark it in accordance with §§45.21 and 45.23 through 45.33, he may apply to the Administrator for a different marking procedure.

45.23 Display of marks; general

(a) Each operator of an aircraft shall display on that aircraft marks consisting of the Roman capital letter "N" (denoting United States region) followed by the Registration number of the aircraft Each suffix letter used in the marks displayed must also be a Roman capital letter.

(b) When marks that include only the Roman capital letter "N" and the registration number are displayed on limited or restricted category aircraft or experimental or provisionally certificated aircraft, the operator shall also display on that aircraft near each entrance to the cabin or cockpit, in letters not less than 2 inches nor more than 6 inches in height, the words "limited," "restricted," "experimental," or "provisional airworthiness," as the case may be.

45.25 Location of marks on fixed-wing aircraft

(a) The operator of a fixed-wing aircraft shall display the required marks on either the vertical tail surfaces or the sides of the fuselage, except as provided in §45.29(f).

(b) The marks required by paragraph (a) of this section shall be displayed as follows:

(1) If displayed on the vertical tail surfaces, horizontally on both surfaces of a single vertical tail or on the outer surfaces of a multivertical tail. However, on aircraft on which marks at least 3 inches high may be displayed in accordance with §45.29 (b)(l), the marks may be displayed vertically on the vertical tail surfaces.

(2) If displayed on the fuselage surfaces, horizontally on both sides of the fuselage between the trailing edge of the wing and the leading edge of the horizontal stabilizer. However, if engine pods or other appurtenances are located in this area and are an integral part of the fuselage side surfaces, the operator may place the marks on those pods or appurtenances.

45.27 Location of marks; nonfixed-wing aircraft

(a) Rotorcraft. Each operator of a rotorcraft shall display on that rotorcraft horizontally on both surfaces of the cabin, fuselage, boom, or tail the marks required by §45.23.

45.29 Size of marks

(a) Except as provided in paragraph (f) of this section, each operator of an aircraft shall display marks on the aircraft meeting the size requirements of this section.

(b) Height. Except as provided in paragraph (h) of this part, the nationality and registration marks must be of equal height and on—

(1) Fixed-wing aircraft, must be at least 12 inches high, except that:

(i) An aircraft displaying marks at least 2 inches high before November 1, 1981, and an aircraft manufactured after November 2, 1981, but before January 1, 1983, may display those marks until the aircraft is repainted or the marks are repainted, restored, or changed;

(ii) Marks at least 3 inches high may be displayed on a glider;

(iii) Marks at least 3 inches high may be displayed on an aircraft for which an experimental certificate has been issued under §21.191 (d) or §21.191 (g) for operating as an exhibition aircraft or as an amateur built aircraft when the maximum cruising speed of the aircraft does not exceed 180 knots.

(iv) Marks may be displayed on an exhibition, antique, or other aircraft in accordance with §45.22.

(2) Airships, spherical balloons, and non-spherical balloons, must be at least 3 inches high; and

(3) Rotorcraft, must be at least 12 inches high, except that rotorcraft displaying before April 18, 1983, marks required by §45.29(b)(3) in effect on April 17, 1983, and rotorcraft manufactured on or after April 18, 1983, but before December 31, 1983, may display those marks until the aircraft is repainted or the marks are repainted, restored, or changed.

(c) Width. Characters must be two-thirds as wide as they are high, except the number "I," which must be one-sixth as wide as it is high, and the letters "M," and "W" which may be as wide as they are high.

(d) Thickness. Characters must be formed by solid lines one-sixth as thick as the character is high.

(e) Spacing. The space between each character may not be less than one-fourth of the character width.

(f) If either one of the surfaces authorized for displaying required marks under §45.25 is large enough for display of marks meeting the size requirements of this section and the other is not, full-size marks shall be placed on the larger surface. If neither surface is large enough for full-size marks, marks as large as practicable shall be displayed on the larger of the two surfaces. If any surface authorized to be marked by §45.27 is not large enough for full-size marks, marks as large as practicable shall be placed on the largest of the authorized surfaces.

(g) Uniformity. The marks required by this Part for fixed-wing aircraft must have the same height, width, thickness, and spacing on both sides of the aircraft.

(h) After March 7, 1988, each operator of an aircraft penetrating an ADIZ or DEWIZ shall display on that aircraft temporary or permanent nationality and registration marks at least 12 inches high.

45.31 Marking of export aircraft

A person who manufactures an aircraft in the United States for delivery outside thereof may display on that aircraft any marks required by the State of registry of the aircraft. However, no person may operate an aircraft so marked within the United States, except for test and demonstration flights for a limited period of time, or while in necessary transit to the purchaser.

Pilots and owners should be very familiar with the contents of their aircraft logbooks. It is the pilot's and owner's responsibility, along with the mechanic's, to ensure that each maintenance entry is legible, accurate, and complete. (See Figs. 8-3 through 8-5.)

March 22, 1994

Total Aircraft Time 1502.0 Hours

Tach Time 972.4 Hours

I certify that this aircraft has been inspected in accordance with an annual inspection as per Air Tractor AT502 owner's manual and was determined to be in an airworthy condition.

Joseph P. Kline

Joseph P. Kline
A&P 123456789 IA

8-3 *Example of maintenance entry.*

March 30, 1994

Aircraft Total Time 1520 Hours

Complied with AD 90-06-03R1, effective date March 27, 1994. Modified the airplane by compliance with paragraph (b) of AD. Installed Cessna Service Kit SK 172-10A. No recurring action required.

Bill Quinlan

Bill Quinlan
A&P 143298671

8-4 *Airworthiness directive compliance logbook entry.*

US Department of Transportation
Federal Aviation Administration

MAJOR REPAIR AND ALTERATION
(Airframe, Powerplant, Propeller, or Appliance)

Form Approved
OMB No. 2120-0020

For FAA Use Only
Office Identification

INSTRUCTIONS: Print or type all entries. See FAR 43.9, FAR 43 Appendix B, and AC 43.9-1 (or subsequent revision thereof) for instructions and disposition of this form. This report is required by law (49 U.S.C. 1421). Failure to report can result in a civil penalty not to exceed $1,000 for each such violation (Section 901 Federal Aviation Act of 1958).

1. Aircraft	Make: Cessna	Model: 182
	Serial No.: 15-10521	Nationality and Registration Mark: N-3763
2. Owner	Name (As shown on registration certificate): William Taylor	Address (As shown on registration certificate): 36 Main Street, Cambria, PA 15946-5555

3. For FAA Use Only

The data identified herein complies with the applicable airworthiness requirements and is approved for the above described aircraft, subject to conformity inspection by a person authorized by FAR Part 43, Section 43.7.

AEA-FSDO-10	April 5, 1994	Ralph Burlingam (signed) Ralph Burlingam
District Office	Date	Signature of FAA Inspector

4. Unit Identification / 5. Type

Unit	Make	Model	Serial No.	Repair	Alteration
AIRFRAME	(As described in Item 1 above)			X	
POWERPLANT					
PROPELLER					
APPLIANCE	Type / Manufacturer				

6. Conformity Statement

A. Agency's Name and Address	B. Kind of Agency	C. Certificate No.
George Morris High Street Johnstown, PA 15236-4444	X U.S. Certificated Mechanic Foreign Certificated Mechanic Certificated Repair Station Manufacturer	1305888

D. I certify that the repair and/or alteration made to the unit(s) identified in item 4 above and described on the reverse or attachments hereto have been made in accordance with the requirements of Part 43 of the U.S. Federal Aviation Regulations and that the information furnished herein is true and correct to the best of my knowledge.

Date	Signature of Authorized Individual
March 19, 1994	George Morris (signed) George Morris

7. Approval for Return To Service

Pursuant to the authority given persons specified below, the unit identified in item 4 was inspected in the manner prescribed by the Administrator of the Federal Aviation Administration and is ☒ APPROVED ☐ REJECTED

BY				
	FAA Flt. Standards Inspector	Manufacturer	X Inspection Authorization	Other (Specify)
	FAA Designee	Repair Station	Person Approved by Transport Canada Airworthiness Group	

Date of Approval or Rejection	Certificate or Designation No.	Signature of Authorized Individual
April 9, 1994	237412	Donald Pauley (signed) Donald Pauley

FAA Form 337 (12-88)

(a)

8-5 *Sample FAA form 337, (a) front and (b) rear.*

NOTICE

Weight and balance or operating limitation changes shall be entered in the appropriate aircraft record. An alteration must be compatible with all previous alterations to assure continued conformity with the applicable airworthiness requirements.

8. Description of Work Accomplished
(If more space is required, attach additional sheets. Identify with aircraft nationality and registration mark and date work completed.)

1. Removed right wing from aircraft and removed skin from outer 6 feet. Repaired buckled spar 49 inches from tip in accordance with attached photographs and figure 1 of drawing dated March 6, 1994.

   DATE: March 15, 1994, inspected splice in Item 1 and found it to be in accordance with data indicated. Splice is okay to cover. Inspected internal and external wing assembly for hidden damage and condition.

   Donald Pauley
   Donald Pauley, A&P 237412 IA

2. Primed interior wing structure and replaced skin P/N's 63-0085, 63-0086, and 63-00878 with same skin 2024-T3, .025 inches thick. Rivet size and spacing all the same as original and using procedures in Chapter 2, Section 3, of AC 43.13-1A, dated 1972.

3. Replaced stringers as required and installed 6 splices as per attached drawing and photographs.

4. Installed wing, rigged aileron, and operationally checked in accordance with manufacturer's maintenance manual.

5. No change in weight or balance.

END

☐ Additional Sheets Are Attached

*U.S. GPO:1989-0-063-171

(b)

9

Aircraft Hardware

Aircraft Fasteners

Aviation fasteners such as nuts, bolts, and rivets are the main means of securing two pieces of material together. Whether the material is aluminum, wood, or even composites, some type of fastener has been engineered and manufactured to perform this important function. Most bolts used in aircraft structures are general purpose Army-Navy (AN), national aircraft standard (NAS), or military standard (MS) hardware.

Aircraft bolts, screws, and nuts are threaded in either the American national coarse (NC) thread series, the American national fine (NF) thread series, the American standard unified coarse (UNC) thread series, or the American standard unified fine (UNF) thread series. One difference between the American national series and the American standard unified series should be pointed out. In the 1-in diameter size, the NF thread specifies 14 threads per inch (1-14NF), whereas the UNF thread specifies 12 threads per inch (1-12UNF). Both threads are designated by the number of times the incline (threads) rotates around a 1-in length of a given diameter bolt or screw. For example, a *4-28* thread indicates that a ¼-in diameter bolt has 28 threads in 1 in of its threaded length.

Threads are also designated by class of fit. The class of a thread indicates the tolerance allowed in manufacturing. Class 1 is a loose fit, class 2 is a free fit, class 3 is a medium fit, and class 4 is a close fit. Aircraft bolts are almost always manufactured in the class 3, medium fit. A class 4 fit requires a wrench to turn the nut onto a bolt, whereas a class 1 fit can easily be turned with the fingers. In general, aircraft screws are manufactured with a class 2 thread fit for ease of assembly.

Bolts and nuts are also produced with right and left threads. A right thread tightens when turned clockwise; a left thread tightens when turned counterclockwise.

Screws

A screw is an externally threaded fastener used to join two parts when great strength is not required. The threaded end of a screw can be blunted or pointed, and usually it has no clearly defined grip length. Several types of structural screws are available that differ from the standard structural bolts only in the type of head. The material is equivalent with a definite grip length. The AN-525 washerhead screws, the AN-509 countersunk structural screws, and the NAS-204 through NAS-235 screws are examples. The material markings are the same as those used on AN standard bolts.

Bolts

Most bolts used in aircraft structures are either the general purpose AN bolts or NAS internal wrenching or close-tolerance bolts. In certain cases, aircraft manufacturers make up special bolts for a particular application.

Identification

The AN aircraft bolts can easily be identified by the code markings located on the bolt heads. (See Fig. 9-1.)

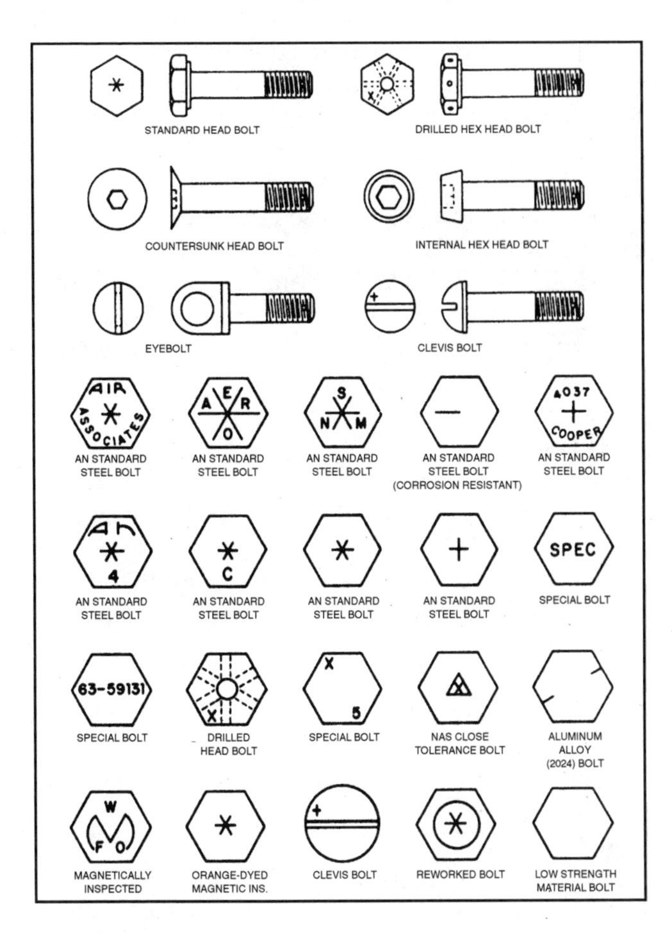

9-1 *Bolt chart.*

These markings generally identify the bolt manufacturer, the type of material used in the manufacturing of the bolt, and if the bolt is a standard AN type or a special-purpose bolt. The AN standard steel bolts are marked with either a raised cross or an asterisk. The corrosion-resistant steel bolt is indicated by a single raised dash, and AN aluminum alloy bolts are marked with two raised dashes. The strength and dimensional details of AN bolts are specified on the Army-Navy aeronautical standard drawings.

Special purpose bolts include the high-strength type, low-strength type, and close tolerance type. Such bolts are normally inspected by magnetic, fluorescent, or equivalent inspection methods. Typical markings include *SPEC*. Close-tolerance NAS bolts are marked with either a raised or recessed triangle. The material markings for NAS bolts are the same as for AN bolts, except that they may be either raised or recessed. Bolts inspected magnetically or by fluorescent means are identified by means of colored lacquer or a head marking of a distinctive type.

Grip Length

In general, the bolt grip lengths should be equal to the material thickness. However, bolts of slightly greater grip length may be used provided washers are placed under the nut or the bolthead. In the case of plate nuts, if proper bolt grip length is not available, add shims under the plate.

Locking or Safetying of Bolts

Lock or safety-tie all bolts and/or nuts, except self-locking nuts. Do not reuse cotter pins and safety wire.

Bolt Fit

Many bolt holes, particularly those in primary connecting elements, have close tolerances. Generally, it is permissible to use the first lettered drill size larger than the normal bolt diameter. In case of oversized or elongated holes in critical structural or support members, consult the manufacturer's structural repair manual, the manufacturer's engineering department, or the Federal Aviation Administration (FAA) guidelines before drilling or reaming the hole to accept the next larger bolt.

Hex-Head Bolts (AN-3 through AN-20)

The hex-head aircraft bolt is an all-purpose structural bolt used for general applications involving tension or shear loads. Alloy steel bolts smaller than no. 10-32 and aluminum alloy bolts smaller than ¼-in diameter are not to be used in primary structure. Do not use aluminum alloy bolts and nuts where they will be repeatedly removed for purposes of maintenance and inspection.

Close-Tolerance Bolts (AN-173–AN 186, Hex-Head; NAS-80–NAS-86)

Close-tolerance bolts are used in high-performance aircraft in applications where the bolted joint is subject to severe load reversals and vibration.

Internal Wrenching Bolts (MS-20004–MS-20024 or NAS495)

These bolts are suitable for use in both tension and shear applications. In steel parts, countersink the bolt hole to seat the large radius of the shank at the head or, as in aluminum alloys, use a special heat-treated washer (NAS-143C) that fits the head to provide adequate bearing area. A special heat-treated plain washer (NAS-143) is used under the nut. Use special high-strength nuts on these bolts.

Drilled-Head Bolts (AN-73)

The AN drilled-head bolt is similar to the standard hex bolt but has a deeper head which is drilled to receive wire for safetying. The AN-3 and the AN-73 series of bolts are interchangeable for all practical purposes from the standpoint of tension and shear strength.

Structural Screws (NAS-204–NAS-235, AN-509 and AN-525)

This type of screw, when made of alloy steel such as SAE4130, NE-8630, or equivalent, and heat-treated from 125,000 lb/in^2, may be used for structural assembly in shear applications similar to structural bolts.

Self-Tapping Screws

The AN-504 and AN 506 screws are used for attaching minor removable parts such as nameplates and the like. AN-530 and AN-531 are used in blind applications for the temporary attachment of sheet metal for riveting and the permanent assembly of nonstructural assemblies. AN-535 is a plain head self-tapping screw used in the attachment of nameplates or in sealing drain holes in corrosion-proofing tubular structures and is not intended to be removed after installation. Never use self-tapping screws to replace standard screws, nuts, bolts, or rivets in the original structure.

Taper Pins (AN-385 AND AN-386)

Plain and threaded taper pins are used in joints which carry shear loads and where absence of play is essential. The plain taper pin is drilled and usually safety-tied with wire. The threaded taper pin is used with a taper pin washer (AN975) and shear nut (safety-tied with cotter pin) or self-locking nut.

The Flathead Pin (MS-20392)

Commonly called a *clevis pin,* the flat-head pin is used in conjunction with tie rod terminals and in secondary controls which are not subject to continuous operation. The pin is customarily installed with the head up so that if the cotter pin falls or works out, the pin will remain in place.

The AN-380 Cotter Pin

This is used for safety-tying bolts, screws, nuts, other pins, and in various applications where such safetying is necessary. Use AN-381 cotter pins in locations where nonmagnetic material is required or where resistance to corrosion is desired.

Nuts

Self-Locking Nuts

Self-locking nuts are acceptable for use on certificated aircraft subject to the restrictions on the pertinent manufacturer's recommended practice sheets. Self-locking nuts are used on aircraft to provide tight connections which will not shake loose under severe vibration. Two types of self-locking nuts are currently in use, the all-metal and the fiber or nylon lock nut. Do not use self-locking nuts at joints which subject either the nut or bolt to rotation. Self-locking nuts may be used with antifriction bearings and control pulleys, provided the inner race of the bearing is clamped to the supporting structure by the nut and bolt. Attach nuts to the structure in a positive manner to eliminate rotation or misalignment when tightening the bolts or screws.

All-metal locknuts are constructed with either the threads in the locking insert out of phase with the load-carrying section or with a saw-cut insert with a pinched-in thread in the locking section. Fiber or nylon locknuts are constructed with an unthreaded fiber-locking insert held securely in place. The fiber or nylon has a smaller diameter than the nut, and when a bolt or screw is entered, it taps into the insert, producing a locking action. After the nut has been tightened, make sure the rounded or chamfered end bolts, studs, or screws extend at least the full round or chamfer through the nut. Flat end bolts, studs, or screws should extend at least $\frac{1}{32}$ in through the nut. When fiber-type self-locking nuts are reused, check the fiber carefully to make sure it has not lost its locking friction or become brittle. Do not reuse locknuts if they can be run up finger tight. Bolts $\frac{5}{16}$-in diameter and over with cotter pin holes may be used with self-locking nuts but only if free from burrs around the holes. Bolts with damaged threads and rough ends are not acceptable.

Self-locking nut bases are made in a number of forms and materials for riveting and welding to aircraft structure or parts. Certain applications require the installation of self-locking nuts in channels, an arrangement which permits the attachment of many nuts with only a few rivets. These channels are track-like bases with regularly spaced nuts which are either removable or nonremovable. The removable type carries a floating nut, which can be snapped in or out of the channel, thus making possible the ready removal of damaged nuts. Nuts such as the clinch and spline, which depend on friction for their anchorage, are not acceptable for use in aircraft structures.

Self-locking nuts may be used on aircraft engines and accessories when their use is specified by the engine manufacturer in bulletins or manuals.

Aircraft Castle Nut (AN-310)

The castle nut is used with drilled-shank AN hex-head bolts, clevis bolts, eye bolts, drilled-head bolts or studs and is designed to accommodate a cotter pin or lockwire as a means of safetying.

Miscellaneous Aircraft Nuts

The *plain nut* (AN-315 and AN-335) has limited use on aircraft structures and requires an auxiliary locking device such as a checknut or lockwasher. *Light hex-nuts* (AN-340 and AN-345) are used in miscellaneous applications and must be locked by an auxiliary device. The *checknut* (AN-316) is used as a locking device for plain nuts, screws, threaded rod ends, and other devices. The *castellated shear nut* (AN-320) is designed for use with clevis bolts and threaded taper pins, which are normally subjected to shearing stress only. Wing nuts (AN-350) are intended for use on hose clamps, battery connections, etc., where the desired tightness is ordinarily obtained by the use of the fingers or hand tools. *Sheet spring nuts,* such as speed nuts, are used with standard and sheet-metal self-tapping screws in nonstructural locations. These nuts have various uses in supporting line clamps, conduit clamps, electrical equipment, access doors, and the like and are available in several types.

Washers

The types of washers used in aircraft structure are plain washers, lockwashers, and special washers.

Plain washers (AN-960 and AN-970) are widely used under hex nuts to provide a smooth bearing surface, to act as a shim, and to adjust holes in bolts. Use plain washers under lockwashers to prevent damage to surfaces. Cadmium-plated steel washers are recommended for use under bolt heads or nuts on aluminum alloy or magnesium structures where corrosion, if it occurs, will then be between the washer and the steel. The AN-970 steel washer provides a greater bearing area than the plain type and is used in wooden structures under both bolt heads and nuts to prevent local crushing of the surface.

Lock washers (AN-935 and AN-936) may be used with machine screws or bolts whenever the self-locking or castellated type of nut is not applicable. Lock washers are not to be used as fastenings to primary or secondary structures or where subject to frequent removal or corrosive conditions.

Ball-socket and seat-washers (AN-950 and AN-955) are used in special applications where the bolt is installed at an angle to the surface or when perfect alignment with the surface is required at all times. These washers are used together.

Rivets

See Fig. 9-2 for a rivet identification chart.

Standard solid-shank rivets and the universal head rivets (AN-470) are used in aircraft construction in both interior and exterior locations. *Roundhead rivets* (AN-430) are used in the interior

Material	Head Marking	AN Material Code	AN425 78° Counter-Sunk Head	AN426 100° Counter-Sunk Head MS20426*	AN427 100° Counter-Sunk Head MS20427*	AN430 Round Head MS20470*	AN435 Round Head MS20613* MS20615*
1100	Plain	A	X	X		X	
2117T	Recessed Dot	AD	X	X		X	
2017T	Raised Dot	D	X	X		X	
2017T-HD	Raised Dot	D	X	X		X	
2024T	Raised Double Dash	DD	X	X		X	
5056T	Raised Cross	B		X		X	
7075-T73	Three Raised Dashes		X	X		X	
Carbon Steel	Recessed Triangle				X		X MS20613*
Corrosion Resistant Steel	Recessed Dash	F			X		X MS20613*
Copper	Plain	C			X		X
Monel	Plain	M			X		
Monel (Nickel-Copper Alloy)	Recessed Double Dots	C					X MS20615*
Brass	Plain						X MS20615*
Titanium	Recessed Large and Small Dot			MS 20426			

* New specifications are for Design purposes

9-2 *Rivet indentification chart.*

of aircraft except where clearance is required for adjacent members. *Flathead rivets* (AN-442) are used in the interior of the aircraft where interference of adjacent members does not permit the use of roundhead rivets. *Brazierhood rivets* (AN-455 and AN-456) are used on the exterior surfaces of aircraft where flush riveting is not essential. All *protruding head rivets* may be replaced by MS-20470 rivets. This has been adopted as the standard for protruding head rivets in this country. *Countersunk head rivets* MS-20426 are used on the exterior surfaces of aircraft to provide a smooth aero-

AN441 Flat Head	AN442 Flat Head MS20470 *	AN455 Brazier Head MS20470 *	AN456 Brazier Head MS20470 *	AN470 Universal Head MS20470 *	* Heat Treat Before Using	Shear Strength P.S.I.	Bearing Strength P.S.I.
	X	X	X	X	No	10000	25000
	X	X	X	X	No	30000	100000
	X	X	X	X	Yes	34000	113000
	X	X	X	X	No	38000	126000
	X	X	X	X	Yes	41000	136000
	X	X	X	X	No	27000	90000
	X	X	X	X	No		
X					No	35000	90000
					No	65000	90000
X					No	23000	
X					No	49000	
					No	49000	
					No		
X					No	95000	

dynamic surface, and in other applications where a smooth finish is desired. The 100-degree countersunk head has been adopted as the standard in this country.

Material Applications

2117-T-4 is the most commonly used rivet material in aluminum alloy structures. Its main advantage lies in the fact that it may be used in the condition received without further treatment. The 2017-T3, 2017-T31, and 2024T4 *rivets* are used in aluminum alloy structures where strength higher than that of the

2117-T4 rivet is needed. The *1100 rivets* of pure aluminum are used for riveting nonstructural parts fabricated from the softer aluminum alloys, such as 1100, 3003, and 5052. When riveting magnesium alloy structures, *5056 rivets* are used exclusively because of their corrosion-resistant qualities in combination with the magnesium alloys.

Mild steel rivets are used primarily in riveting steel parts. Do not use galvanized rivets on steel parts subjected to high heat. *Corrosion-resistant steel rivets* are used primarily in riveting corrosion-resistant steel parts such as firewalls, exhaust stack bracket attachments, and similar structures. *Monel rivets* are used in special cases for riveting high-nickel steel alloys and nickel alloys. They may be used interchangeably with stainless steel rivets as they are more easily driven. However, it is preferable to use stainless steel rivets in stainless steel parts.

Copper rivets are used for riveting copper alloys, leather, and other nonmetallic materials. This rivet has only limited usage in aircraft.

High-shear rivets are sometimes used in connections where the shearing loads are the primary design consideration, and their use is restricted to such connections. Note that high-shear rivet patterns are not to be used for the installation of control surface hinges and hinge brackets. Do not paint the rivets prior to assembly, even where dissimilar metals are being joined. However, it is advisable to touch up each end of the driven rivet with zinc chromate primer to allow the later application of the general airplane finish.

Blind rivets in the MS-20600 through MS20603 series rivets and the mechanically locked stem NAS 1398, 1399, 1738, and 1739 rivets may be substituted for solid rivets in accordance with the blind rivet or aircraft manufacturer's recommendations. Blind rivets should not be used where the looseness or failure of a few rivets will impair the airworthiness of the aircraft.

AN-type aircraft solid rivets can be identified by code markings on the rivet heads. A rivet made of I 100 material is designated as an *A* rivet and has no head marking. The 2217-T4 rivets are designated *AD* rivets and have a dimple on the head. The 2017-T4 alloy rivets are designated as *D* rivets and have a raised teat on the head. Two dashes on a rivet head indicate a 2024-T4 alloy designated as *DD*. A *B* designation is given to a rivet of 5056-H-12 material and is marked with a raised cross on the rivet head.

Fasteners (Cowl and Fairing)

A number of patented fasteners are in use on aircraft. A variety of these fasteners are commercially available, and the manufacturer's recommendations concerning the proper use of these types of fasteners should always be considered in other than replacement application.

Torques

The use of the correct torque cannot be overemphasized. Undertorque can result in unnecessary wear of nuts and bolts as well as the parts they are holding together. When insufficient pressures are applied, uneven loads will be transmitted throughout the assembly, which may result in excessive wear or premature failure caused by fatigue. Overtorque can be equally damaging, because the bolt or nut may fail from overstressing the threaded areas. There are a few simple, but very important, procedures that should be followed to ensure that correct torque is applied:

1. Calibrate the torque wrench periodically to ensure accuracy; recheck frequently.
2. Be sure that bolt and nut threads are clean and dry (unless otherwise specified by the manufacturer).

Table 9-1. Standard Torque Table

Bolt, stud, or screw size		Torque values in inch-pounds for tightening nuts			
		On standard bolts, studs, and screws having a tensile strength of 125,000–140,000 lb/in^2		On bolts, studs, and screws having a tensile strength of 140,000–160,000 lb/in^2	On high-strength bolts, studs, and screws having a tensile strength 160,000 lb/in^2 and over
		Shear type nuts (AN-320, AN-364, or equivalent)	Tension type nuts and threaded machine parts (AN-310, AN-365, or equivalent)	Any nut, except shear type	Any nut, except shear type
8–32	8–36	7–9	12–15	14–17	15–18
10–24	10–32	12–15	20–25	23–30	25–35
¼–20		25–30	40–50	45–49	50–68

	1/4–28	30–40	50–70	60–80	70–90
5/16–18		48–55	80–90	85–117	90–144
	5/16–24	60–85	100–140	120–172	140–203
3/8–16		95–110	160–185	173–217	185–248
	3/8–24	95–110	160–190	175–271	190–351
7/16–14		140–155	235–255	245–342	255–428
	7/16–20	270–300	450–500	475–628	500–756
1/2–13		240–290	400–480	440–636	480–792
	1/2–20	290–410	480–690	585–840	690–990
9/16–12		300–420	500–700	600–845	700–990
	9/16–18	480–600	800–1,000	900–1,220	1,000–1,440
5/8–11		420–540	700–900	800–1,125	900–1,350
	5/8–18	660–780	1,100–1,300	1,200–1,730	1,300–2,160
3/4–10		700–950	1,150–1,600	1,380–1,925	1,600–2,250

(Cont.)

Table 9-1. (*Continued*)

Bolt, stud, or screw size		Torque values in inch-pounds for tightening nuts			
		On standard bolts, studs, and screws having a tensile strength of 125,000–140,000 lb/in^2		On bolts, studs, and screws having a tensile strength of 140,000–160,000 lb/in^2	On high-strength bolts, studs, and screws having a tensile strength 160,000 lb/in^2 and over
		Shear type nuts (AN320, AN364, or equivalent)	Tension type nuts and threaded machine parts (AN-310, AN365, or equivalent)	Any nut, except shear type	Any nut, except shear type
	3/4–16	1,300–1,500	2,300–2,500	2,400–3,500	2,500–4,500
7/8–9		1,300–1,800	2,200–3,000	2,600–3,570	3,000–4,140
	7/8–14	1,500–1,800	2,500–3,000	2,750–4,650	3,000–6,300

1″–8		2,200–3,000	3,700–5,000	4,350–5,920	5,000–6,840
	1″–14	2,200–3,300	3,700–5,500	4,600–7,250	5,500–9,000
1⅛–8		3,300–4,000	5,500–6,500	6,000–8,650	6,500–10,800
	1⅛–12	3,000–4,200	5,000–7,000	6,000–10,250	7,000–13,500
1¼–8		4,000–5,000	6,500–8,000	7,250–11,000	8,000–14,000
	1¼–12	5,400–6,600	9,000–11,000	10,000–16,750	11,000–22,500

SOURCE: F.A.A. AC 43.13.

3. Run nut down by hand to near contact with the washer or bearing surface.
4. Use one torque wrench to achieve the desired torque recommended by the manufacturer, or obtain desired torque as shown in Table 9-1.
5. Apply a smooth even pull when applying torque pressure.
6. When installing a castle nut, start alignment with the cotter pin hole at minimum recommended torque. If the hole and nut castellation do not align, change washers and try again. Exceeding the maximum recommended torque is not recommended.
7. If torque is applied to capscrews or bolt heads, apply recommended torque plus friction drag torque.
8. If special adapters are used which will change the effective length of the torque wrench, the final torque indication or wrench setting must be adjusted accordingly. Determine the torque wrench indication or setting with adapter installed as shown in Table 9-1. Table 9-1 is a composite chart of recommended torque to be used when specific torque is not recommended by the manufacturer. The chart includes standard nut and bolt combinations currently used in aviation maintenance.

Owner Maintenance

Replacement of Nonstructural Fasteners

Replacing fasteners or screws requires using the exact replacement fastener as the one being replaced. The best procedure is to match the fastener being installed with the exact same type and size of the fastener being replaced. Do not attempt to replace any fastener that is structural in nature. If unsure of the

9-3 *Use the correct size screwdriver to remove screws.*

function of a fastener, be sure to check with a mechanic as to its function.

Many pilots want to replace the myriad of screws on their aircraft with stainless steel. There are numerous screw kits available through aircraft parts suppliers to accomplish this task. Although the job sounds easy, it is time-consuming.

To replace nonstructural fasteners, use a screwdriver and follow these steps:

1. The most difficult part of this job is matching the removed screws with replacement screws. Work in one area at a time. Remove the screws with a screwdriver (see Fig. 9-3).
2. find the exact size replacement screw in the kit and replace. Continue this process until all the screws have been replaced.
3. Make the appropriate logbook entry. (See Chap. 8.)

Safety Wire

Owner-accomplished preventive maintenance frequently requires the removal and consequently the replacement of safety wire. It's a simple task but one that does require understanding of the numerous methods of safety wiring. Remember the principal requirement for safety wire is to create tension which in turn tends to the tightening of the bolt, screw, nut, or fastener.

Using safety wire pliers or smooth-jawed duckbill pliers and aviation-grade safety wire, follow these steps:

1. Insert a length of safety wire through the hole in the fastener. Match the two ends of wire together and twist together by hand or with duck bill pliers. If available a special tool designed to install and twist safety wire should be used. Do not twist the wire together too tightly! A tight twist will seriously weaken the safety wire.
2. Adjust the length of the twisted portion of wire so that it fits between the two fasteners (see Fig. 9-4).

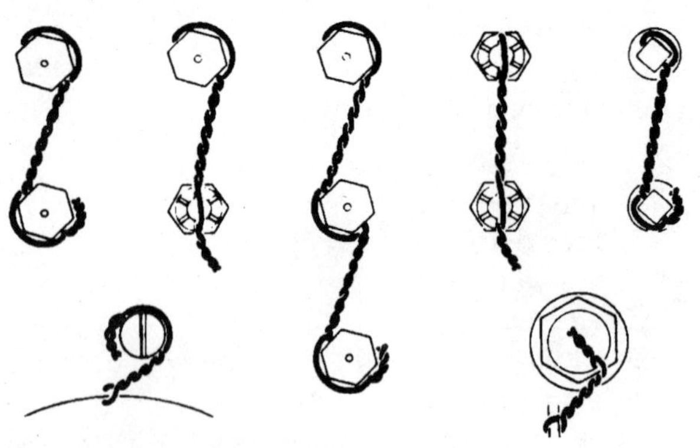

9-4 *Examples of safety wire.*

3. Be sure that the wire is situated on the correct side of the fastener and exerts a tightening pull on the fastener. Insert one end of the wire into the hole of the next fastener to be safety wired and again twist the ends together.
4. Cut off the excess wire and bend the stub back toward the fastener.
5. Complete the logbook entry. (See Chap. 8.)

10

Tool Care and Proper Use

This chapter contains information on some common hand tools used in aviation repair work. It outlines the basic knowledge required in using the most common hand tools and measuring instruments. The use of tools may vary, but good practices for safety, care, and storage of tools remain the same.

General-Purpose Tools

Hammers and Mallets

Figure 10-1 shows some of the hammers that the aviation mechanic may be required to use. Metal-head hammers are sized according to the weight of the head without the handle.

Occasionally it is necessary to use a soft-faced hammer, which has a striking surface made of wood, brass, lead, rawhide, hard rubber, or plastic. These hammers are intended for use in forming soft metals and striking surfaces that are easily damaged. Soft-faced hammers should not be used for normal hammering. Striking nails will quickly ruin this hammer.

A mallet is a hammerlike tool with a head made of hickory, rawhide, or rubber. The mallet is handy for shaping thin metal parts without denting them.

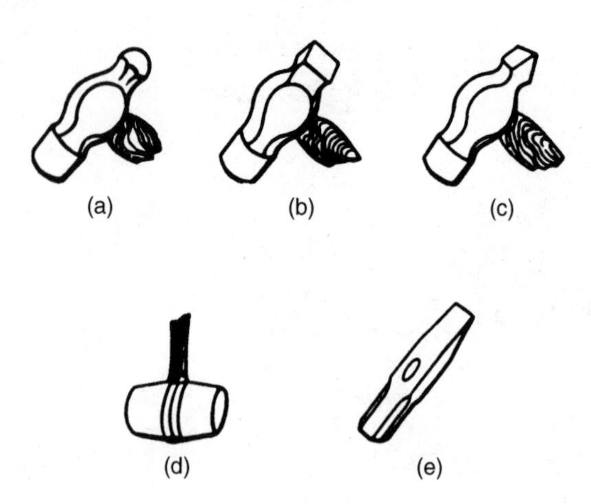

10-1 *Hammers. (a) Ball peen. (b) Straight peen. (c) Cross peen. (d) Tinner's mallet. (e) Riveting hammer.*

Always use a wooden mallet when pounding a wood chisel or a gouge.

Choose the correct hammer or mallet for the job. Ensure that the handle is tight. When striking a blow with the hammer, use the forearm as an extension of the handle. Swing the hammer by bending the elbow, not the wrist. Always strike the work squarely with the full face of the hammer. Always keep the faces of hammers and mallets smooth and free from dents to prevent marring the work.

Screwdrivers

The screwdriver can be classified by its shape, type of blade, and blade length. It is made for only one purpose, that is, for loosening or tightening screws or screwhead bolts. Figure 10-2 shows several different types of screwdrivers. When using the common screwdriver, select the largest screwdriver whose blade will make a good fit in the screw which is to be turned.

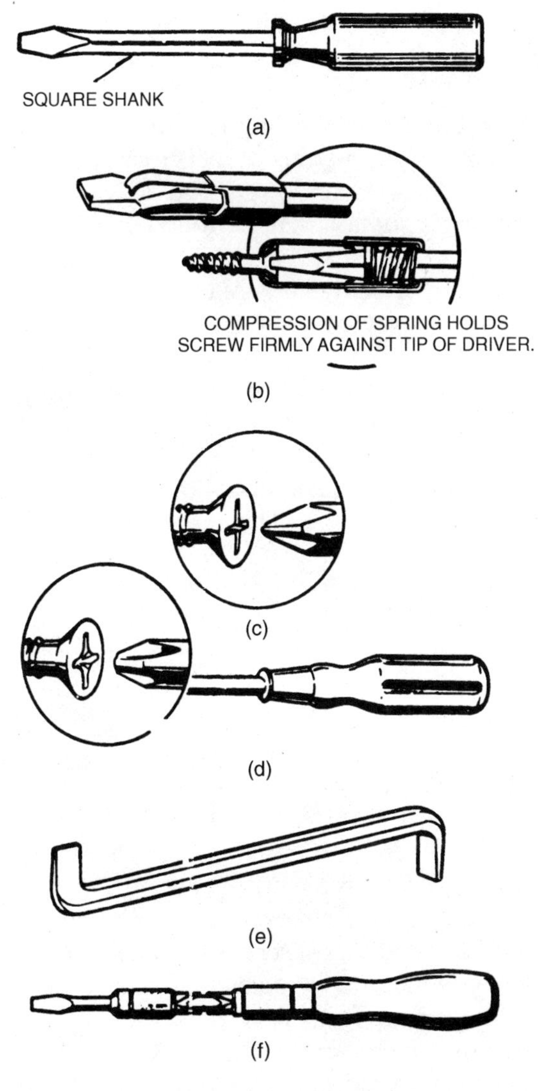

10-2 *Screwdrivers. (a) Common type. (b) Screw-holding driver. (c) Reed and Prince driver. (d) Phillips head driver. (e) Offset driver. (f) Ratchet and spiral driver.*

A common screwdriver must fill at least 75 percent of the screw slot. If the screwdriver is the wrong size, it cuts and burrs the screw slot, making it worthless. A screwdriver with the wrong size blade may slip and damage adjacent parts of the structures.

The common screwdriver is used only where slotted head screws or fasteners are found on aircraft. An example of a fastener which requires the use of a common screwdriver is the airlock fastener which is used to secure the cowling on some aircraft.

The two types of recessed head screws in common use are the Phillips and the Reed and Prince. Both the Phillips and Reed and Prince recessed heads are optional on several types of screws. As shown in Fig. 10-2*c*, the Reed and Prince recessed head forms a perfect cross. The screwdriver used with this screw is pointed on the end. Because the Phillips screw has a slightly larger center in the cross, the Phillips screwdriver is blunt on the end. The Phillips screwdriver is not interchangeable with the Reed and Prince. The use of the wrong screwdriver results in mutilation of the screwdriver and the screwhead. When turning a recessed head screw, use only the proper recessed head screwdriver of the correct size.

An offset screwdriver may be used when vertical space is limited. Offset screwdrivers are constructed with both ends bent 90 degrees to the shank handle. By using alternate ends, most screws can be seated or loosened even when the swinging space is limited. Offset screwdrivers are made for common, both standard and recessed head screws.

A screwdriver should not be used for chiseling or prying. Do not use a screwdriver to check an electric circuit, since an electric arc will burn the tip and make it useless. In some cases, an electric arc may fuse the blade to the unit being checked.

When using a screwdriver on a small part, always hold the part in tire vise or rest it on a workbench.

Do not hold the part in the hand, as the screwdriver may slip and cause serious personal injury.

The ratchet or spiral screwdriver is fast acting in that it turns the screw when the handle is pulled back and then pushed forward. It can be set to turn the screw either clockwise or counterclockwise, or it can be locked in position and used as a standard screwdriver. The ratchet screwdriver is not a heavy-duty tool and should be used only for light work. A word of caution: When using a spiral or ratchet screwdriver, extreme care must be used to maintain constant pressure and prevent the blade from slipping out from the slot in the screw head. If this occurs, the surrounding structure is subject to damage.

Pliers and Plier-Type Cutting Tools

There are several types of pliers, but those used most frequently in aircraft repair work are the diagonal, adjustable combination, needlenose, and duckbill. The size of pliers indicates their overall length, usually ranging from 5 to 12 in.

The 6-in slip-joint plier is the preferred size for use in repair work. The slip joint permits the jaws to be opened wider at the hinge for gripping objects with large diameters. Slip-joint pliers come in sizes from 5 to 10 in. The better grades are drop-forged steel.

Flatnose pliers are very satisfactory for making flanges. The jaws are square, fairly deep, and usually well matched, and the hinge is firm. These are characteristics which give a sharp, neat bend.

Roundnose pliers are used to crimp metal. They are not made for heavy work because too much pressure will spring the jaws, which are often wrapped to prevent scarring the metal.

Needlenose pliers have half-round jaws of varying lengths. They are used to hold objects and make adjustments in tight places.

Duckbill pliers resemble a "duck's bill" in that the jaws are thin, flat, and shaped like a duck's bill. They are used exclusively for twisting safety wire.

Water pump (channel locks) pliers are slip-joint pliers with the jaws set at an angle to the handles. The most popular type has the slip-joint channeled; hence the name channel locks. These are used to grasp packing nuts, pipe, and odd-shaped parts.

Diagonal pliers are usually referred to as *diagonals* or *dikes*. The diagonal is a short-jawed cutter with a blade set at a slight angle on each jaw. This tool can be used to cut wire, rivets, small screws, and cotter pins. besides being practically indispensable in removing or installing safety wire. The duckbill pliers and the diagonal cutting pliers are used extensively in aviation for the job of safety wiring.

Two important rules for using pliers are:

1. Do not make pliers work beyond their capacity. The longnosed variety are especially delicate. It is easy to spring or break them or nick the edges. If this occurs, they are practically useless.
2. Do not use pliers to turn nuts. In just a few seconds, a pair of pliers can damage a nut more than years of service.

Punches

Punches are used to locate centers for drawing circles, to start holes for drilling, to punch holes in sheet metal, to transfer location of holes in patterns, and to remove damaged rivets, pins, or bolts.

Solid or hollow punches are the two types generally used. Solid punches arc classified according to the shape of their points. Figure 10-3 shows several types of punches.

Prick punches are used to place reference marks on metal. This punch is often used to transfer dimen-

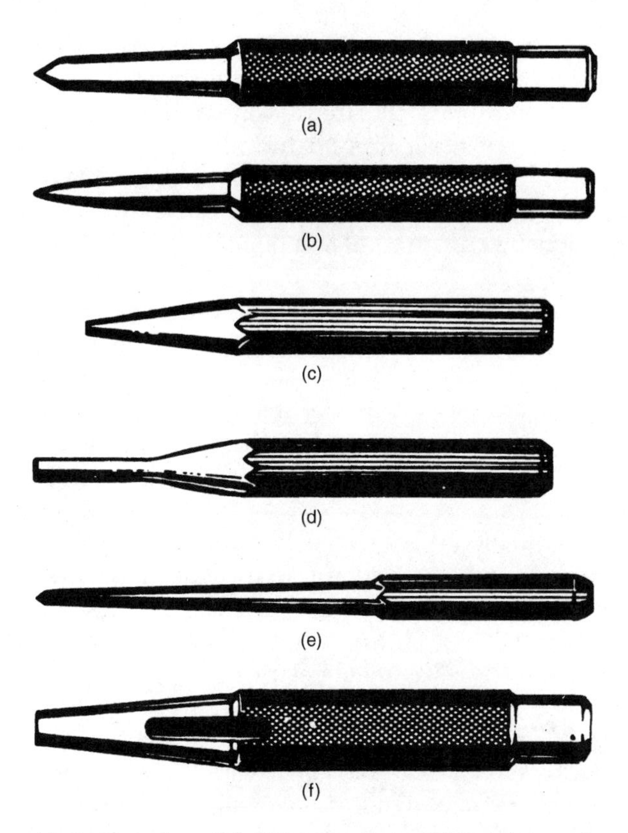

10-3 *Punches. (a) Center punch. (b) Prick punch. (c) Starting punch. (d) Pin punch. (e) Aligning punch. (f) Hollow shank gasket punch.*

sions from a paper pattern directly on the metal. To do this, first place the paper pattern directly on the metal. Then go over the outline of the pattern with the prick punch, tapping it lightly with a small hammer and making slight indentations on the metal at the major points on the drawing. These indentations can then be used as reference marks for cutting the metal. A prick punch should never be struck a heavy blow with a hammer because doing so may bend the punch or cause excessive damage to the material being worked.

Large indentations in metal, needed to start a twist drill, are made with a center punch. The center punch should never be struck with enough force to dimple the material around the indentation or to cause the metal to protrude through the other side of the sheet. A center punch has a heavier body than a prick punch and is ground to a point with an angle of about 60 degrees.

The drive punch, which is often called a *tapered punch,* is used for driving out damaged rivets, pins, and bolts which sometimes bind in holes. The drive punch is therefore made with a flat face instead of a point. The size of the punch is determined by the width of the face, which is usually $\frac{1}{8}$ to $\frac{1}{4}$ in.

Pin punches, often called *drift punches,* are similar to drive punches and are used for the same purposes. The difference in the two is that the sides of a drive punch taper all the way to the face, whereas the pin punch has a straight shank. Pin punches are sized by the diameter of the face, in thirty-seconds of an inch, and range from $\frac{1}{16}$ to $\frac{3}{8}$ in in diameter.In general practice, a pin or bolt which is to be driven out is usually started and driven with a drive punch until the sides of the punch touch the side of the hole. A pin punch is then used to drive the pin or bolt the rest of the way out of the hole. Stubborn pins may be started by placing a thin piece of scrap copper, brass, or aluminum directly against the pin and then striking it with a hammer until the pin begins to move.

Never use a prick punch or center punch to remove objects from holes, because the point of the punch will spread the object and cause it to bind even more.

The transfer punch is usually about 4 in long. It has a point that tapers, then turns straight for a short distance in order to fit a drill-locating hole in a template. The tip has a point similar to that of a prick punch. As its name implies, the transfer punch is used to transfer the location of holes through the template or pattern to the material.

Wrenches

The wrenches most often used in aircraft maintenance are classified as open-end, box-end, socket, adjustable, and special wrenches. The alien wrench, although seldom used, is required on one special type of recessed screw. One of the most widely used metals for making wrenches is chrome vanadium steel. Wrenches made of this metal are almost unbreakable.

Solid, nonadjustable wrenches with open parallel jaws on one or both ends are known as *open-end wrenches*. These wrenches may have their jaws parallel to the handle or at an angle up to 90 degrees; most are set at an angle of 15 degrees. Basically, the wrenches are designed to fit a nut, bolt head, or other object to allow a turning action.

Box-end wrenches are popular tools because of their usefulness in close quarters. They are called *box-end wrenches* since they box, or completely surround, the nut or bolt head. Practically all box-end wrenches are made with 12 points so they can be used in places with as little as 15 degrees swing.

Although box-end wrenches are ideal to break loose tight nuts or pull tight nuts tighter, time is lost turning the nut off the bolt once the nut is broken loose. Only when there is sufficient clearance to rotate the wrench in a complete circle can this tedious process be avoided.

After a tight nut is broken loose, it can be completely backed off or unscrewed more quickly with an open-end than with a box-end wrench. In this case, a combination wrench is needed, which has a box-end on one end and an open-end wrench of the same size on the other. Both the box-end and combination wrenches are shown in Fig. 10-4.

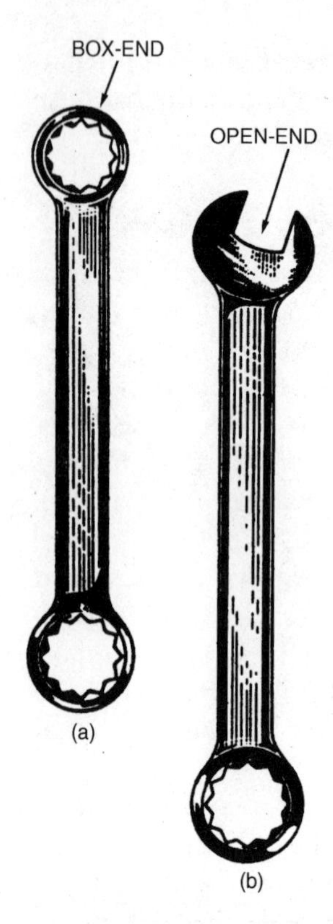

10-4 *Wrenches. (a) Box-end wrench. (b) Combination wrench.*

A socket wrench is made of two parts: (1) The socket, which is placed over the top of a nut or bolt head, and (2) a handle, which is attached to the socket. Many types of handles, extensions, and attachments are available to make it possible to use socket wrenches in almost any location or position. Sockets arc made with either fixed or detachable handles. Socket wrenches with fixed handles are usually furnished as an accessory to a machine. They have a 4-, 6-, or 12-sided recess to fit a nut or bolt head that needs regular adjustment.

Sockets with detachable handles usually come in sets and fit several types of handles, such as the T ratchet, screwdriver grip, and speed handle. Socket wrench handles have a square lug on one end that fits into a square recess in the socket head. The two parts are held together by a light spring-loaded poppet. Two types of sockets, a set of handles, and an extension bar are shown in Fig. 10-5.

The adjustable wrench is a handy utility tool which has smooth jaws and is designed as an open-end wrench. One jaw is fixed, but the other may be moved by a thumbscrew or spiral screwworm adjustment in the handle. The width of the jaws may be varied from 0 to ½ in or more. The angle of the opening to the handle is 22½ degrees on an

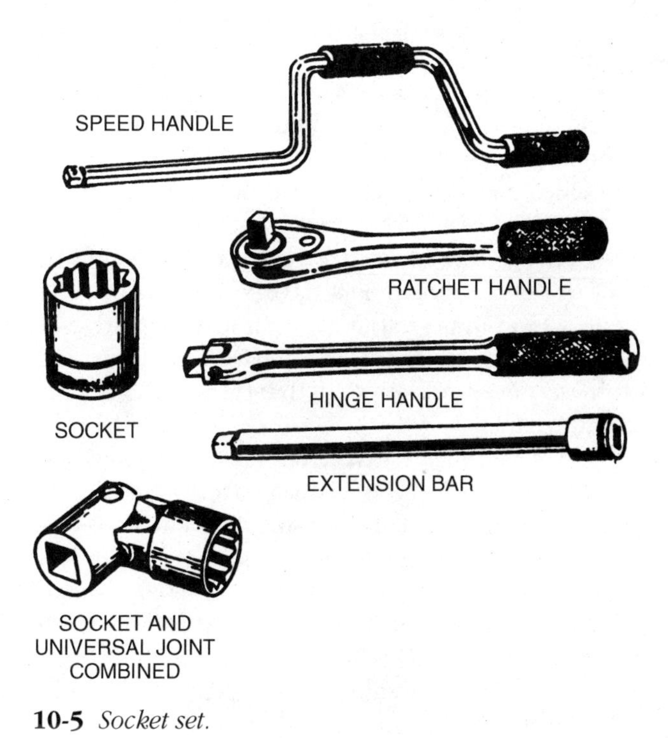

10-5 *Socket set.*

adjustable wrench. One adjustable wrench does the work of several open-end wrenches. Although versatile, they are not intended to replace the standard open-end, box-end, or socket wrenches. When using any adjustable wrench, always exert the pull on the side of the handle attached to the fixed jaw of the wrench.

Special Wrenches

The category of special wrenches includes the spanner, torque, and alien wrenches. The hook spanner is for a round nut with a series of notches cut in the outer edge. This wrench has a curved arm with a hook on the end which fits into one of the notches on the nut. The hook is placed in one of these notches with the handle pointing in the direction the nut is to be turned.

Some hook spanner wrenches are adjustable and will fit nuts of various diameters. U-shaped hook spanners have two lugs on the face of the wrench to fit notches cut in the face of the nut or screw plug. End spanners resemble a socket wrench but have a series of lugs that fit into corresponding notches in a nut or plug. Pin spanners have a pin in place of a lug, and the pin fits into a round hole in the edge of a nut. Face-pin spanners are similar to the U-shaped hook spanners except that the former have pins instead of lugs.

There are times when definite pressure must he applied to a nut or bolt. In such cases a torque wrench must be used. The torque wrench is a precision tool consisting of a torque indicating handle and appropriate adapter or attachments. This wrench measures the amount of turning or twisting force applied to a nut, bolt, or screw.

The deflecting beam, micrometer-setting, and dial-indicating types are the three most commonly used torque wrenches. When using the deflecting beam and the dial-indicating torque wrenches, the torque is read visually on a dial or scale mounted on the handle of the wrench. The micrometer-setting torque wrench is preset to the desired torque. When this torque is reached, a sharp impulse or breakaway is noticed by the operator.

Before each use, the torque wrench should be visually inspected for damage. If a bent pointer, cracked or broken glass (dial type), or signs of rough handling are found, the wrench must be tested. Torque wrenches must be tested at periodic intervals to ensure accuracy.

Most headless setscrews are the alien type and must be installed and removed with an alien wrench. Allen wrenches are six-sided bars in the shape of an L. They range in size from $\frac{3}{64}$ to $\frac{1}{2}$ in and fit into a hexagonal recess in the setscrew.

Metal-Cutting Tools

Hand Snips

There are several kinds of hand snips, each of which serves a different purpose. Straight, curved, hawks-bill, and aviation snips are in common use (see Fig. 10-6). Straight snips are used for cutting straight lines when the distance is not great enough to use a squaring shear and for cutting the outside of a curve. The other types are used for cutting the inside of curves or radii. Snips should never be used to cut heavy sheet metal.

Aviation snips are designed especially for cutting heat-treated aluminum alloy and stainless steel. They are also adaptable for enlarging small holes. The blades have small teeth on the cutting edges and are shaped for cutting very small circles and irregular outlines. The handles are the compound leverage type, making it possible to cut material as thick as 0.051 in. Aviation snips are available in two types,

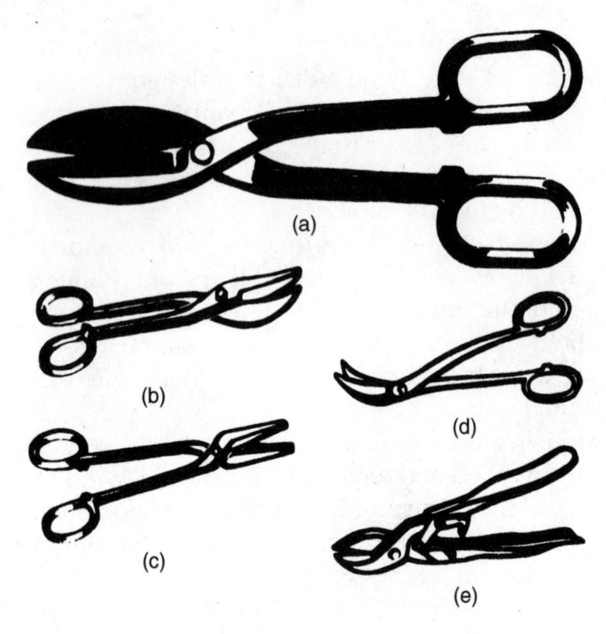

10-6 *Cutting snips. (a) Straight hand snips. (b) Circle snips. (c) Trojan snips. (d) Hawks-bill snips. (e) Aviation snips.*

those which cut from right to left and those which cut from left to right.

Unlike the hacksaw, snips do not remove any material when the cut is made, but minute fractures often occur along the cut. Therefore, cuts should be made about $\frac{1}{32}$ in from the layout line and finished by hand-filing down to the line.

Hacksaws

The common hacksaw has a blade, a frame, and a handle. The handle can be obtained in two styles, pistol grip and straight (see Fig. 10-7).

Hacksaw blades have holes in both ends; they are mounted on pins attached to the frame. When installing a blade in a hacksaw frame, mount the blade with the teeth pointing forward, away from the handle.

Blades are made of high-grade tool steel or tungsten steel and are available in sizes from 6 to 16 in in length. The 10-in blade is most commonly used. There are two types, the all-hard blade and the flexible blade. In flexible blades, only the teeth are hardened. Selection of the best blade for the job involves finding the right type and pitch. An all-hard blade is best for sawing brass, tool steel, cast iron, and heavy cross-section materials. A flexible blade is usually best for sawing hollow shapes and metals having a thin cross-section.

The pitch of a blade indicates the number of teeth per inch. Pitches of 14, 18, 24, and 32 teeth per inch are available. A blade with 14 teeth per inch is preferred when cutting machine steel, cold-rolled steel, or structural steel. A blade with 18 teeth per inch is preferred for solid stock aluminum, bearing

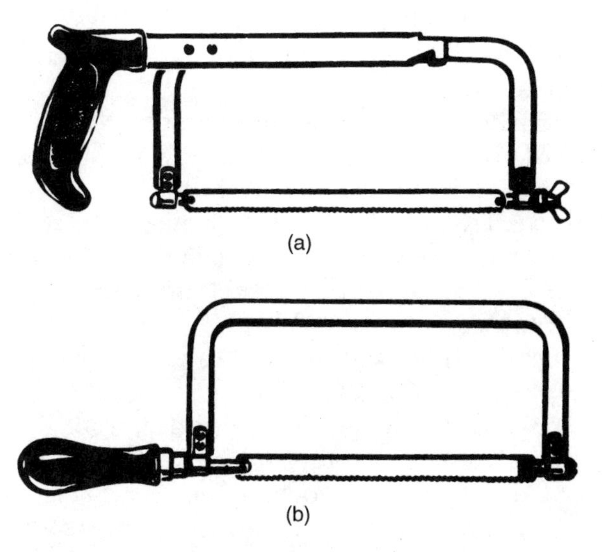

10-7 *Hacksaws. (a) Pistol grip. (b) Straight grip.*

metal, tool steel, and cast iron. Use a blade with 24 teeth per inch when cutting thick-walled tubing, pipe, brass, copper, channel, and angle iron. Use a 32-teeth-per-inch blade for cutting thin-walled tubing and sheet metal.

When using a hacksaw, observe the following procedures:

1. Select an appropriate saw blade for the job.
2. Assemble the blade in the frame so that the cutting edge of the teeth points away from the handle.
3. Adjust tension of the blade in the frame to prevent the saw from buckling and drifting.
4. Clamp the work in the vise in such a way that will provide as much bearing surface as possible and will engage the greatest number of teeth.
5. Indicate the starting point by nicking the surface with the edge of a file to break any sharp comer that might strip the teeth. This mark will also aid in starting the saw at the proper place.
6. Hold the saw at an angle that will keep at least two teeth in contact with the work at all times. Start the cut with a light, steady, forward stroke just outside the cutting line. At the end of the stroke, relieve the pressure and draw the blade back. (The cut is made on the forward stroke.)
7. After the first few strokes, make each stroke as long as the hacksaw frame will allow. This will prevent the blade from overheating. Apply just enough pressure on the forward stroke to cause each tooth to remove a small amount of metal. The strokes should be long and steady with a speed not more than 40 to 50 strokes per minute.

8. After completing the cut, remove chips from the blade, loosen tension on the blade, and return the hacksaw to its proper place.

Chisels

A chisel is a hard steel–cutting tool which can be used for cutting and chipping any metal softer than the chisel itself. It can be used in restricted areas and for such work as shearing rivets or splitting seized or damaged nuts from bolts (see Fig. 10-8).

The size of a flat cold chisel is determined by the width of the cutting edge. Length will vary, but chisels are seldom under 5 in or over 8 in long.

Chisels are usually made of eight-sided tool steel bar stock. Since the cutting edge is convex, the center portion receives the greatest shock when cutting, and the weaker corners are protected. The cutting angle should be 60 to 70 degrees for general use, such as for cutting wire, strap iron, or small bars and rods.

When using a chisel, hold it firmly in one hand. With the other hand, strike the chisel head squarely with a ball-peen hammer.

When cutting square corners or slots, a special cold chisel called a cape chisel should be used. It is like a flat chisel except the cutting edge is very narrow. It has the same cutting angle and is held and used in the same manner as any other chisel.

Rounded or semicircular grooves and corners which have fillets should be cut with a roundness chisel. This chisel is also used to recenter a drill which has moved away from its intended center.

The diamond point chisel is tapered square at the cutting end, then ground at an angle to provide the sharp diamond point. It is used for cutting B grooves and inside sharp angles.

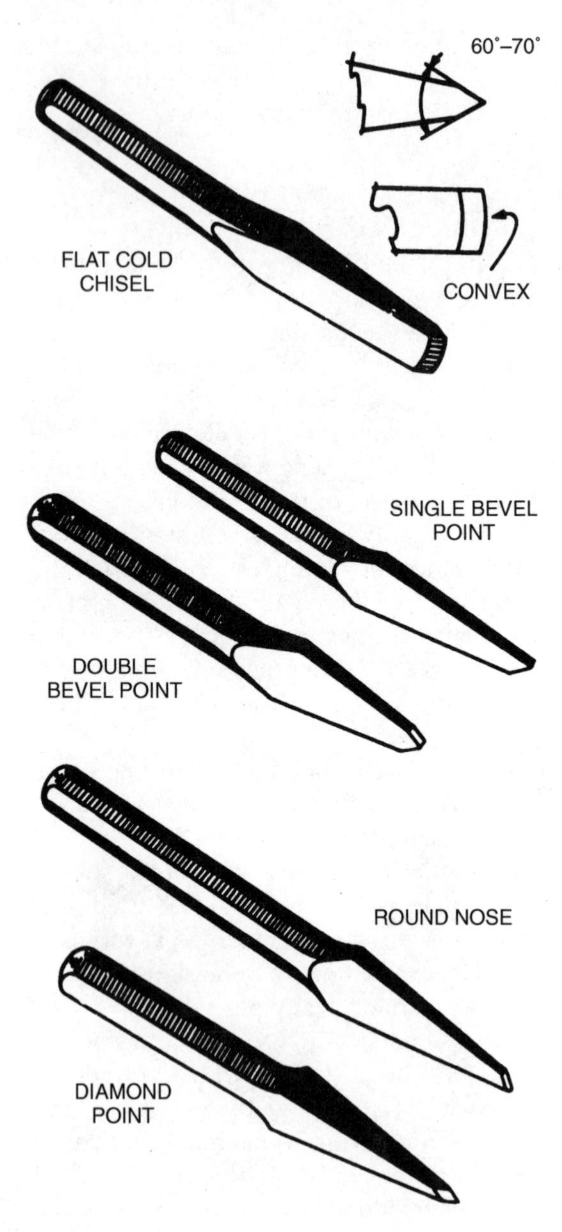
60°–70°
FLAT COLD
CHISEL
CONVEX
SINGLE BEVEL
POINT
DOUBLE
BEVEL POINT
ROUND NOSE
DIAMOND
POINT

10-8 *Chisels.*

Files

Most files are made of high-grade tool steels that are hardened and tempered. Files are manufactured in a variety of shapes and sizes. They are known either by the cross-section, the general shape, or by their particular use. The cuts of files must be considered when selecting them for various types of work and materials.

Files are used to square ends, file rounded comers, remove bars and slivers from metal, straighten uneven edges, file holes and slots, and smooth rough edges.

Files have three distinguishing features: (1) Their length, measured exclusive of the tang (see Fig. 10-9); (2) their kind or name, which has reference to the relative coarseness of the teeth; and (3) their cut.

Files are usually made in two types of cuts, single-cut and double-cut. The single-cut file hag a single row of teeth extending across the face at an angle of 65 to 85 degrees with the length of the file. The size of the cuts depends on the coarseness of the file. The double-cut file has two rows of teeth which cross each other. For general work, the angle of the first row is 40 to 45 degrees. The first row is generally referred to as *overcut* and the second row as *upcut*; the upcut is somewhat finer and not so deep as the overcut.

Files — Care and Use

Files and rasps are cataloged in three ways:

Length. Measuring from the tip to the heel of the file. The tang is never included in the length.

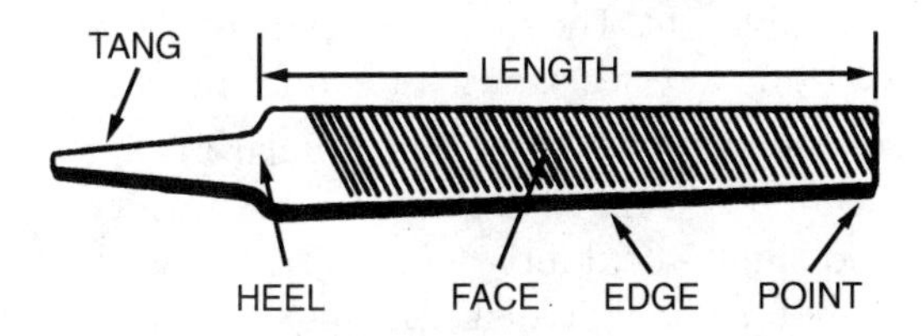

10-9 *Files.*

Shape. Refers to the physical configuration of the file (circular, rectangular, or triangular or a variation thereof).

Cut. Refers to both the character of the teeth or the coarseness; rough, coarse, and bastard for use on heavier classes of work and second cut, smooth, and dead smooth for finishing work.

Most Commonly Used Files

The following files are used most often:

Hand files. These are parallel in width and tapered in thickness. They have one safe edge (smooth edge) which permits filing in corners and on other work where a safe edge is required. Hand files are double-cut and used principally for finishing flat surfaces and similar work.

Flat files. These files are slightly tapered toward the point in both width and thickness. They cut on both edges as well as on the sides. They are the most common files in use. Flat files are double-cut on both sides and single-cut on both edges.

Mill files. These are usually tapered slightly in thickness and in width for about one-third of their length. The teeth are ordinarily single-cut. These files are used for draw-filing and to some extent for filing soft metals.

Square files. These files may be tapered or blunt and are double-cut. They are used principally for filing slots and key seats, and for surface filing.

Round or rattail files. These are circular in cross-section and may be either tapered or blunt and single- or double-cut. They are used principally for filing circular openings or concave surfaces.

Triangular and three-square files. These files are triangular in cross-section. Triangular files are single-cut and are used for filing the gullet between saw teeth. Three-square files, which are double-cut, may be used for filing internal angles, clearing out corners, and filing taps and cutters.

Half-round files. These files cut on both the flat and round sides. They may be single- or double-cut. Their shape permits them to be used where other files would be unsatisfactory.

Lead float files. These are especially designed for use on soft metals. They are single-cut and are made in various lengths.

Warding file. Rectangular in section and tapered to narrow point as to width, these files are used for narrow space filing where other files cannot be used.

Knife file. A knife-blade section is used by tool and die makers on work having acute angles.

Wood file. Similar to flat and half-round files, a wood file has coarser teeth and is especially adaptable for use on wood.

Vixen (curved-tooth) files. Curved tooth files are especially designed for rapid filing and smooth finish on soft metals and wood. The regular cut is adapted for tough work on cast iron, soft steel, copper, brass, aluminum, wood, slate, marble, fiber, rubber, and the like. The fine cut gives excellent results on steel, cast iron, phosphor bronze, white brass, and all hard metals. The smooth cut is used where the amount of material to be removed is very slight but where a super finish is desired.

The following methods are recommended for using files:

1. *Crossfiling*. Before attempting to use a file, place a handle on the tang of the file. This is essential for proper guiding and safe use. In moving the file endwise across the work (commonly known as *cross-filing*), grasp the handle so that its end fits into and against the fleshy part of the palm with the thumb lying along the top of the handle in a lengthwise direction. Grasp the end of the file between the thumb and first two fingers. To prevent undue wear, relieve the pressure during the return stroke.
2. *Draw-filing*. A file is sometimes used by grasping it at each end, crosswise to the work, then moving it lengthwise with the work. When done properly, work may be finished somewhat finer than when cross-filing with the same file. In draw-filing, the teeth of the file produce a shearing effect. To accomplish this shearing effect, the angle at which the file is held with respect to its line of movement varies with different files, depending on the angle at which the teeth are cut. Pressure should be relieved during the back stroke.
3. *Rounding corners*. The method used in filing a rounded surface depends upon its width and the radius of the rounded surface. If the surface is narrow or only a portion of a surface is to be rounded, start the forward stroke of the file with the point of the file inclined downward at approximately a 45-degree angle. Using a rocking chair motion, finish the stroke with the heel of the file near the curved surface. This method allows use of the full length of the file.
4. *Removing burred or slivered edges*. Practically every cutting operation on sheet metal produces burrs or slivers. These must

be removed to avoid personal injury and to prevent scratching and marring of parts to be assembled. Burrs and slivers will prevent parts from fitting properly and should always be removed from the work as a matter of habit.

Lathe filing requires that the file be held against the work revolving in the lathe. The file should not be held rigid or stationary but should be stroked constantly with a slight gliding or lateral motion, along the work. A standard mill file may be used for this operation, but the long angle lathe file provides a much cleaner shearing and self-clearing action. Use a file with "safe" edges to protect work with shoulders from being marred.

Care of Files

There are several precautions that any good craftsperson will take in caring for files:

1. Choose the right file for the material and work to be performed.
2. Keep all files racked and separated so they do not bear against each other.
3. Keep the files in a dry place—rust will corrode the teeth points.
4. Keep files clean—tap the end of the file against the bench after every few strokes, to loosen and clear the filings. Use the file card to keep files clean—a dirty file is a dull file.

Particles of metal collect between the teeth of a file and may make deep scratches in the material being filed. When these particles of metal are lodged too firmly between the teeth and cannot be removed by tapping the edge of the file, remove them with a file card or wire brush (see Fig. 10-10). Draw the brush across the file so that the bristles pass down the gullet between the teeth.

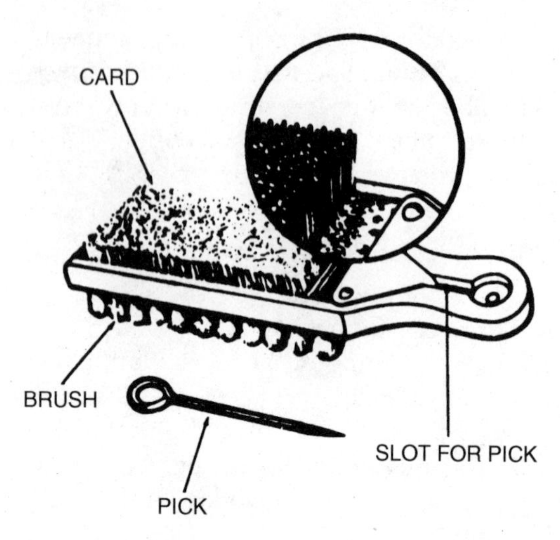

10-10 *File card.*

Drills

There are generally four types of portable drills used in aviation for holding and turning twist drills. Holes ¼ in in diameter and under can be drilled using a hand drill. This drill is commonly called an *egg beater*. The breast drill is designed to hold larger size twist drills than the hand drill. In addition a breast plate is affixed at the upper end of the drill to permit the use of body weight to increase the cutting power of the drill. Electric and pneumatic power drills are available in various shapes and sizes to satisfy almost any requirement. Pneumatic drills are preferred for use around flammable materials, since sparks from an electric drill are a fire or explosion hazard.

A twist drill is a pointed tool that is rotated to cut holes in material. It is made of a cylindrical hardened steel bar having spiral flutes (grooves) running the length of the body and a conical point with cutting edges formed by the ends of the flutes.

Twist drills are made of carbon steel or high-speed alloy steel. Carbon steel twist drills are satisfactory for the general run of work and are relatively inexpensive. The more expensive high-speed twist drills are used for the tough materials such as stainless steels. Twist drills have from one to four spiral flutes. Drills with two flutes are used for most drilling; those with three or four flutes are used principally to follow smaller drills or to enlarge holes.

The principal parts of a twist drill are the shank, the body, and the point, illustrated in Fig. 10-11. The drill shank is the end that fits into the chuck of a hand or power drill. The two shank shapes most commonly used in hand drills are the straight shank and the square or bit stock shank (see Fig. 10-11). The straight shank generally is used in hand, breast, and portable electric drills; the square shank is made to fit into a carpenter's brace. Tapered shanks generally are used in machine shop drill presses.

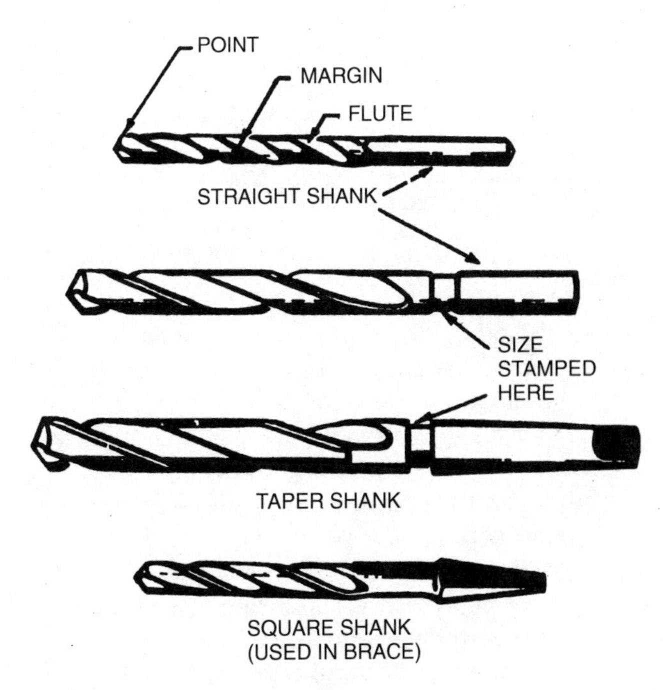

10-11 *Drills.*

The metal column forming the core of the drill is the body. The body clearance area lies just back of the margin, slightly smaller in diameter than the margin, to reduce the friction between the drill and the sides of the hole. The angle at which the drill point is ground is the lip clearance angle. On standard drills used to cut steel and cast iron, the angle should be 59 degrees from the axis of the drill. For faster drilling of soft materials, sharper angles are used.

The diameter of a twist drill may be given in one of three ways: by (1) fractions, (2) letters, or (3) numbers. Fractionally, they are classified by sixteenths of an inch (from 1/16 to 3½ in), by thirty-seconds (from 1/32 to 2½ in), or by sixty-fourths (from 1/64 to 1¼ in). For a more exact measurement a letter system is used with decimal equivalents, *A* (0.234 in) to Z (0.413 in). The number system of classification is most accurate: 80 (0.0314 in) to 1 (0.228 in).

The twist drill should be sharpened at the first sign of dullness. For most drilling, a twist drill with a cutting angle of 118 degrees (59 degrees on either side of center) will be sufficient; however, when drilling soft metals, a cutting angle of 90 degrees may be more efficient.

Typical procedures for sharpening drills are as follows:

1. Adjust the grinder to rest at a convenient height for resting the back of the hand while grinding.
2. Hold the drill between the thumb and index finger of the right or left hand. Grasp the body of the drill near the shank with the other hand.
3. Place the band on the tool rest with the center line of the drill, making a 59 degree angle with the cutting face of the grinding wheel. Lower the shank end of the drill slightly.

4. Slowly place the cutting edge of the drill against the grinding wheel. Gradually lower the shank of the drill as you twist the drill in a clockwise direction. Maintain pressure against the grinding surface only until you reach the heel of the drill.
5. Check the results of grinding with a gauge to determine whether the lips are the same length and at a 59 degree angle.

Reamers

Reamers are used to smooth and enlarge holes to exact size. Hand reamers have square end shanks so that they can be turned with a tap wrench or similar handle.

A hole that is to be reamed to exact size must be drilled about 0.003- to 0.007-in undersize. A cut that removes more than 0.007-in places too much load on the reamer and should not be attempted.

Reamers are made of either carbon tool steel or high-speed steel. The cutting blades of a high-speed steel reamer lose their original keenness sooner than those of a carbon steel reamer; however, after the first super-keenness is gone, they are still serviceable. The high-speed reamer usually lasts much longer than the carbon steel type.

Reamer blades are hardened to the point of being brittle and must be handled carefully to avoid chipping. When reaming a hole, rotate the reamer in the cutting direction only. Turn the reamer steadily and evenly to prevent chattering or marking and scoring of the hole walls.

Reamers are available in any standard size. The straight-fluted reamer is less expensive than the spiral-fluted reamer, but the spiral type has less tendency to chatter. Both types are tapered for a short distance back of the end to aid in starting. Bottoming reamers have no taper and are used to complete the reaming of blind holes.

For general use, an expansion reamer is the most practical. This type is furnished in standard sizes from ¼ in to 1 in, increasing in diameter by $\frac{1}{32}$-in increments.

Taper reamers, both hand- and machine operated, are used to smooth and true tapered holes and recesses.

Countersinks

A countersink is a tool which cuts a cone-shaped depression around the hole to allow a rivet or screw to get flush with the surface of the material. Countersinks are made with various angles to correspond to the various angles of the countersunk rivet and screwheads.

Special stop countersinks are available. Stop countersinks are adjustable to any desired depth, and the cutters are interchangeable so that holes of various countersink angles may be made. Some stop countersinks have a micrometer set arrangement (in increments of 0.001 in) for adjusting the cutting depths.

When using a countersink, be careful not to remove an excessive amount of material, since this reduces the strength of flush joints.

Layout and Measuring Tools

Layout and measuring devices are precision tools. They are carefully machined, accurately marked, and, in many cases, made up of very delicate parts. When using these tools, be careful not to drop, bend, or scratch them. The finished product will be no more accurate than the measurements or the layout; therefore, it is very important to understand how to read, use, and care for these tools.

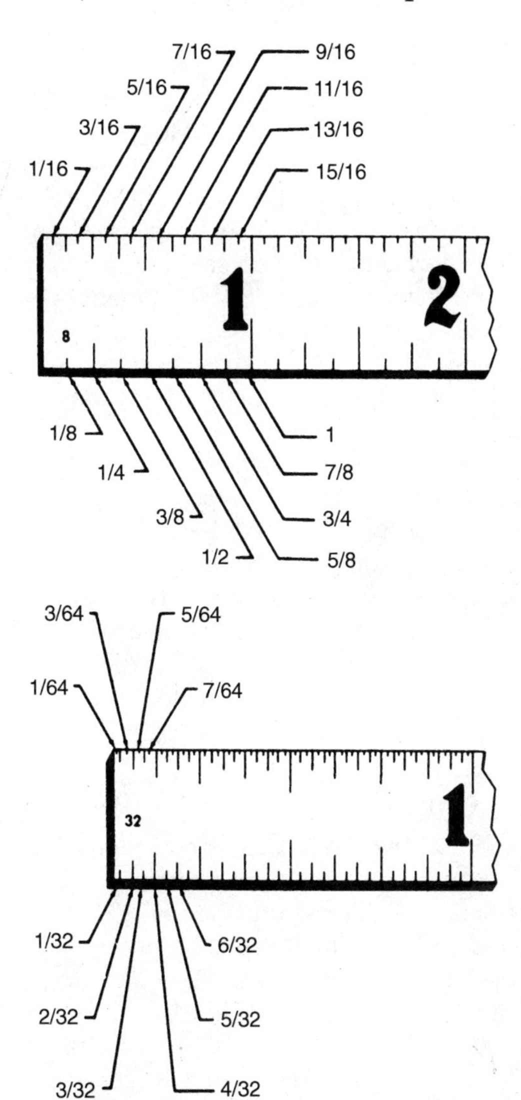
7/16
9/16
5/16
11/16
3/16
13/16
1/16
15/16
8
1
2
1/8
1
1/4
7/8
3/8
3/4
1/2
5/8
3/64
5/64
1/64
7/64
32
1
1/32
6/32
2/32
5/32
3/32
4/32

10-12 *Rules.*

Rules

Rules are made of steel and are either rigid or flexible. The flexible steel rule will bend, but it should not be bent intentionally as it may break rather easily.

In aircraft work the unit of measure most commonly used is the inch. The inch may be divided into smaller parts by means of either common or decimal fraction divisions. The fractional divisions for an inch are found by dividing the inch into equal parts—halves ($1/2$), quarters ($1/4$), eighths ($1/8$), sixteenths ($1/16$), thirty-seconds ($1/32$), and sixty-fourths ($1/64$)— as shown in Fig. 10-12.

The fractions of an inch may be expressed in decimals, called *decimal equivalents* of an inch; for example, $1/8$ in is expressed as 0.0125 (one hundred twenty-five ten-thousandths of an inch).

Rules are manufactured in two basic styles, those divided or marked in *common* fractions (Fig. 10-12) and those divided or marked in decimals or divisions of $1/100$th of an inch. A rule may be used either as a measuring tool or as a straightedge.

Combination Sets

The combination set (Fig. 10-13), as its name implies, is a tool that has several uses. It can be used for the same purposes as an ordinary tri-square, but it differs from the tri-square in that the head slides along the blade and can he clamped at any desired place. Combined with the square or stock head are a level and scriber. The head slides in a central groove on the blade or scale, which can be used separately as a rule.

The spirit level in the stock head makes it convenient to square a piece of material with a surface and at the same time tell whether one or the other is plumb or level. The head can be used alone as a simple level.

The combination of square head and blade can also be used as a marking gauge to scribe lines at a 45-degree angle, as a depth gauge, or as a height gauge.

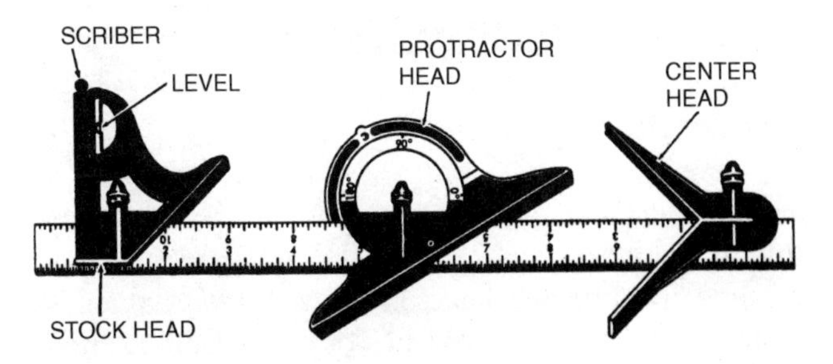

10-13 *Combination set.*

A convenient scriber is held frictionally in the head by a small brass bushing.

The center head is used to find the center of shafts or other cylindrical work. The protractor head can be used to check angles and also may be set at any desired angle to draw lines.

Scriber

The scriber is designed to serve the aviation mechanic in the same way a pencil or pen serves a writer. In general, it is used to scribe or mark lines on metal surfaces. The scriber (Fig. 10-14) is made of tool steel, 4 to 12 in long, and has two needle-pointed ends. One end is bent at a 90-degree angle for reaching and marking through holes.

Before using a scriber always inspects the points for sharpness. Be sure the straightedge is flat on the metal and in position for scribing. Tilt the scriber slightly in the direction toward which it will be moved, holding it like a pencil. Keep the scriber's point close to the guiding edge of the straightedge. The scribed line should be heavy enough to be visible but no deeper than necessary to serve its purpose.

Dividers and Pencil Compasses

Dividers and pencil compasses have two legs joined at the top by a pivot. They are used to scribe circles

10-14 *Scriber.*

and area and for transferring measurements from the rule to the work.

Pencil compasses have one leg tapered to a needle point; the other leg has a pencil or pencil lead inserted. Dividers have both legs tapered to needle points.

When using pencil compasses or dividers, the following procedures are suggested:

1. Inspect the points to make sure they are sharp.
2. To set the dividers or compasses, hold them with the point of one leg in the graduations on the rule. Turn the adjustment nut with the thumb and forefinger; adjust the dividers or compasses until the point of the other leg rests on the graduation of the rule which gives the required measurement.
3. To draw an arc or circle with either the pencil compasses or dividers, hold the thumb attachment on the top with the thumb and forefinger. With pressure exerted on both legs, swing the compass in a clockwise direction and draw the desired arc or circle.
4. The tendency for the legs to slip is avoided by inclining the compasses or dividers in the direction in which they are being rotated. In working on metals, the dividers are used only to scribe arcs or circles that will later be removed by cutting. All other arcs or circles are drawn with pencil compasses to avoid scratching the material.
5. On paper layouts, the pencil compasses are used for describing arcs and circles. Dividers should be used to transfer critical measurements because they are more accurate than a pencil compass.

Calipers

Calipers are used for measuring diameters and distances or for comparing distances and sizes. The three common types of calipers are the inside, the outside, and the hermaphrodite calipers, such as gear-tool calipers. (See Fig. 10-15.)

Outside calipers are used for measuring outside dimensions, for example, the diameter of a piece of round stock. Inside calipers have outward curved legs for measuring inside diameters, such as diameters of holes, the distance between two surfaces, the width of slots, and other similar jobs. A hermaphrodite caliper is generally used as a marking gauge in layout work. It should not be used for precision measurement.

Micrometer Calipers

There are four types of micrometer calipers, each designed for a specific use. The four types are commonly called *outside micrometer, inside micrometer, depth micrometer,* and *thread micrometer.* Micrometers

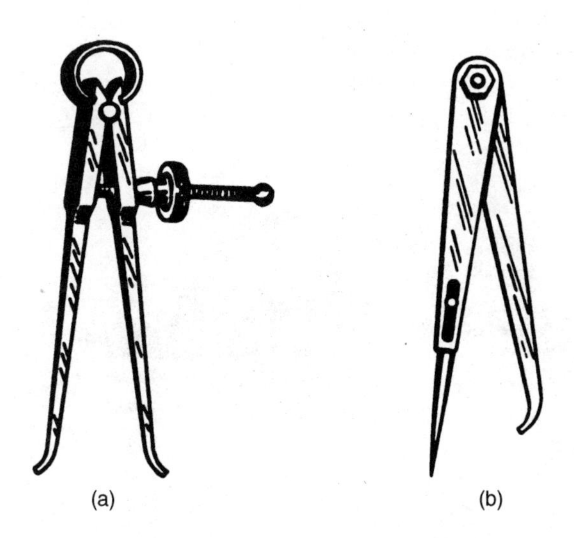

10-15 *Calipers. (a) Spring inside calipers. (b) Hermaphrodite calipers.*

are available in a variety of sizes, in sizes of 0 to ½ in, 0 to 1 in, 1 to 2 in, 2 to 3 in, 3 to 4 in, 4 to 5 in, or 5 to 6 in.

The outside micrometer (Fig. 10-16) is used by the mechanic more often than any other type. It may be used to measure the outside dimensions of shafts, thickness of sheet metal stock, diameter of drills, and for many other applications.

The smallest measurement which can be made with the use of the steel role is $\frac{1}{64}$ in common fractions, and $\frac{1}{100}$ in in decimal fractions. To measure more closely than this (in thousandths and ten-thousandths of an inch), a micrometer is used. If a dimension given in a common fraction is to be measured with the micrometer, the fraction must be converted to its decimal equivalent.

All four types of micrometers are read in the same way. The method of reading an outside micrometer is discussed later. Fixed parts of a micrometer (Fig. 10-16) are the frame, barrel, and anvil. The movable parts of a micrometer are the thimble and spindle. The thimble rotates the spindle, which moves in the threaded portion inside the barrel. Turning the thimble provides an opening between the anvil and the end of the spindle where the work is measured. The size of

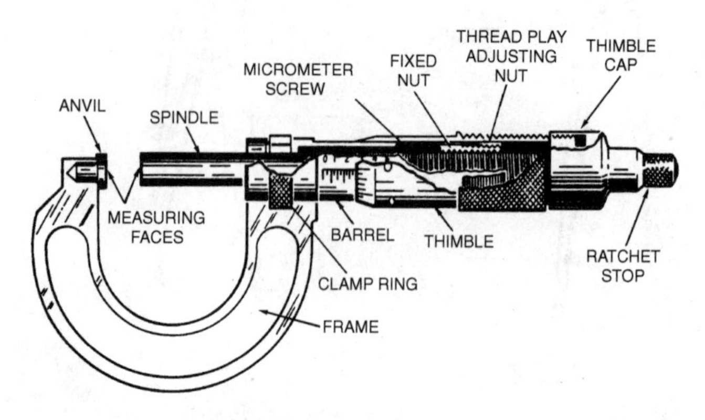

10-16 *Outside micrometer.*

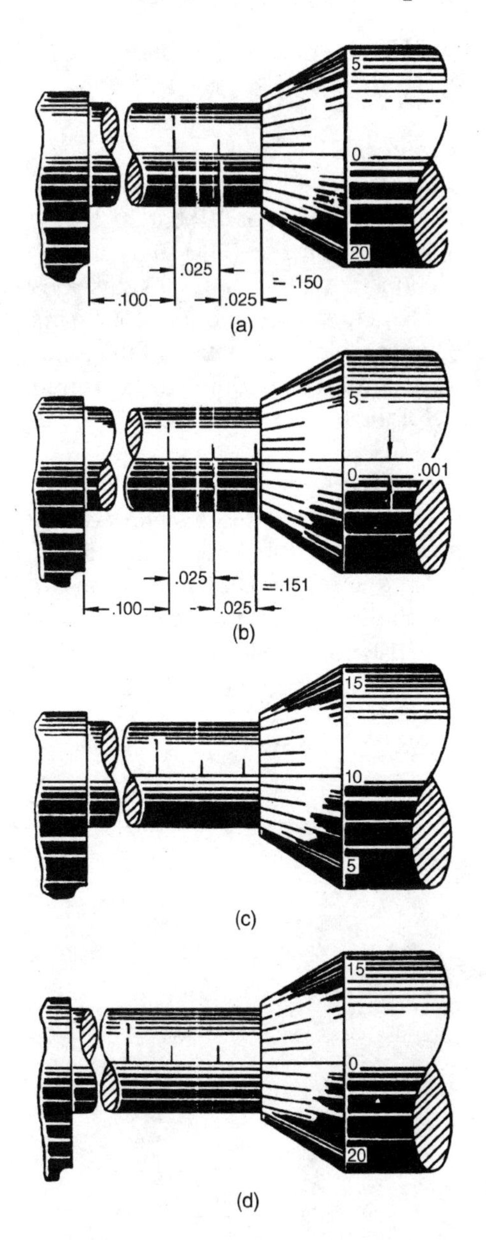

10-17 *Reading a micrometer. (a) 0.5 in. (b) 0.151 in. (c) 0.16 in. (d) 0.175 in.*

the work is indicated by the graduations on the barrel and thimble.

Reading a Micrometer.
The lines on the barrel marked 1, 2, 3, 4, and so on, indicate measurements of tenths, or 0.100 in, 0.200 in, 0.300 in, 0.400 in, respectively (see Fig. 10-17).

Each of the sections between the tenths divisions (between, 0.2, 0.3, 0.4, etc.) is divided into four parts of 0.025 in each. One complete revolution of the thimble (from zero on the thimble around to the same zero) moves it one of these divisions (0.025 in) along the barrel.

The bevel edge of the thimble is divided into 25 equal parts. Each of these parts represents one twenty-fifth of the distance the thimble travels along the barrel in moving from one of the 0.025-in divisions to another. Thus, each division on the thimble represents 1⁄1000 (0.001) of an inch. These divisions are marked for convenience at every five spaces by 0. 5, 10, 15.

Index

About the Author

Douglas S. Carmody (Beaufort, SC) is the author of the McGraw-Hill Pilot Test Guide Series. He is an airline captain with USAirways (formerly USAir), a flight instructor, an A&P mechanic, and operator of a flight training school specializing in airplane owner/pilot education.